ENCYCLOPAEDIA OF LIFE SCIENCES & ENVIRONMENTAL SCIENCES

ENCYCLOPAEDIA OF
LIFE SCIENCES &
ENVIRONMENTAL SCIENCES

Vinaya Sahay

ANMOL PUBLICATIONS PVT. LTD.
NEW DELHI - 110 002 (INDIA)

ANMOL PUBLICATIONS PVT. LTD.
H.O.: 4374/4B, Ansari Road, Daryaganj,
New Delhi-110 002 (India)
Ph.: 23278000, 23261597
B.O.: No. 1015, Ist Main Road, BSK IIIrd Stage
IIIrd Phase, IIIrd Block,
Bangalore - 560 085 (India)
Visit us at: www.anmolpublications.com

Encyclopaedia of Life Sciences & Environmental Sciences

ISBN 978-81-261-3594-3

PRINTED IN INDIA

Printed at Mehra Offset Press, Delhi.

Contents

Preface

This publication titled "Encyclopaedia of Life Sciences & Environmental Sciences" is an initiative to deal with the subject in its totality. Life Sciences is the study of life and is concerned with such topics as the characteristics, classification, and behaviors of organisms, how species come into existence, and the interactions they have with each other and with the natural environment. It is a science that deals with the functions or problems of living organisms. Environmental science is the study of interactions among physical, chemical, and biological components of the environment. It is an interdisciplinary science overlapping the categories in Natural sciences, Engineering sciences and Social sciences. In nature, Environmental science focuses on pollution and degradation of the environment related to human activities and their impact on biodiversity and sustainability. As an interdisciplinary field, environmental science also applies knowledge from economics, law and social sciences. It can be applied to cyberspace environment also.

Environmental science came alive as a substantive, active field of scientific investigation in the 1960s and 1970s. This was driven by the need for a large multi-disciplined team to analyze complex environmental problems, the arrival of substantive environmental laws requiring specific environmental protocols of investigation, and growing public awareness of a need for action in addressing environmental problems. Environmental science encompasses issues such as climate change, conservation, biodiversity, groundwater and soil contamination, use of natural resources, waste management, sustainable development, air pollution and noise pollution. Due to the interdisciplinary nature of environmental science, teams of professionals commonly work together to conduct environmental research or to produce Environmental Impact Statements, as required by the U.S. National Environmental Policy Act (NEPA) or under state laws. Other professional organizations engender work in environmental science and aid in communication among the diverse sciences. Today, all these are a rapidly growing field of research and development that is cutting across many traditional boundaries. This encyclopaedia titled "Encyclopaedia of Life Sciences & Environmental Science" is really unique in terms of coverage, scope, elaboration, usefulness and futuristic perspective. It will provide readers with a user-friendly approach towards acquainting themselves with various select traditional and newly emerging fields of biosciences. This encyclopaedia is of great contemporary relevance. It will serve as a reference book on the said subject and tends to acquaint readers with both in-depth and extensive knowledge of the subject concerned. This encyclopaedia is a unique initiative to provide the most comprehensive coverage to the subject area in question. This encyclopedia mainly focuses on the following areas: Biology & Life Sciences; Ecology and Environmental Science; Atmospheric Science and Earth Science; Energy Development and Renewable Energy; Green Building, Green Computing & Other Technologies; Recycling, Waste and Wastewater Treatment and Management etc.

—Editor

1

Introduction to Biology & Life Science

BIOLOGY

Biology, also referred to as the biological sciences, is the scientific study of life. Biology examines the structure, function, growth, origin, evolution, and distribution of living things. It classifies and describes organisms, their functions, how species come into existence, and the interactions they have with each other and with the natural environment. Four unifying principles form the foundation of modern biology: cell theory, evolution, genetics and homeostasis.

Biology as a separate science was developed in the nineteenth century, as scientists discovered that organisms shared fundamental characteristics. Biology is now a standard subject of instruction at schools and universities around the world, and over a million papers are published annually in a wide array of biology and medicine journals.

Most biological sciences are specialized disciplines. Traditionally, they are grouped by the type of organism being studied: botany, the study of plants; zoology, the study of animals; and microbiology, the study of microorganisms. The fields within biology are further divided based on the scale at which organisms are studied and the methods used to study them: biochemistry examines the fundamental chemistry of life; molecular biology studies the complex interactions of systems of biological molecules; cellular biology examines the basic building block of all life, the cell; physiology examines the physical and chemical functions of the tissues and organ systems of an organism; and ecology examines how various organisms and their environment interrelate.

Foundations of Modern Biology

There are four unifying principles of biology:

- Cell theory. All living organisms are made of at least one cell, the basic unit of function in all organisms. In addition, the core mechanisms and chemistry of all cells in all organisms are similar, and cells emerge only from preexisting cells that multiply through cell division.
- Evolution. Through natural selection and genetic drift, a population's inherited traits change from generation to generation.
- Gene theory. A living organism's traits are encoded in their DNA, the fundamental component of genes. In addition, traits are passed on from one generation to the next by way of these genes. All information flows from genes to the phenotype, the observable physical or biochemical characteristics of the organism. Although the phenotype expressed by the gene may adapt to the environment of the organism, that information is not transferred back to the genes. Only through the process of evolution do genes change in response to the environment.
- Homeostasis. The physiological processes that allow an organism to maintain its internal environment notwithstanding its external environment.

Cell Theory

The cell is the fundamental unit of life. Cell theory states that all living things are composed of one or more cells, or the secreted products of those cells, for example, shell and bone. Cells arise from other cells through cell division, and in multicellular organisms, every cell in the organism's body is produced from a single cell in a fertilized egg. Furthermore, the cell is considered to be the basic part of the pathological processes of an organism.

Evolution

A central organizing concept in biology is that life changes and develops through evolution and that all lifeforms known have a common origin (see Common descent). This has led to the striking similarity of units and processes discussed in the previous section. Charles Darwin established evolution as a viable theory by articulating its driving force, natural selection (Alfred Russel Wallace is recognized as the co-discoverer of this concept). Darwin theorized that species and breeds developed through the processes of natural selection as well as by artificial selection or selective breeding. Genetic drift was embraced as an additional mechanism of evolutionary development in the modern synthesis of the theory.

The evolutionary history of the species— which describes the characteristics of the various species from which it descended— together with its genealogical relationship to every other species is called its phylogeny. Widely varied approaches to biology generate information about phylogeny. These include the comparisons of DNA sequences conducted within molecular biology or genomics, and comparisons of fossils or other records of ancient organisms in paleontology. Biologists organize and analyze evolutionary relationships through various methods, including phylogenetics, phenetics, and cladistics. For a summary of major events in the evolution of life as currently understood by biologists, see evolutionary timeline.

Up into the 19th century, it was commonly believed that life forms could appear spontaneously under certain conditions (see spontaneous generation). This misconception was challenged by William Harvey's diction that "all life [is] from [an] egg" (from the Latin "Omne vivum ex ovo"), a foundational concept of modern biology. It simply means that there is an unbroken continuity of life from its initial origin to the present time.

A group of organisms shares a common descent if they share a common ancestor. All organisms on the Earth both living and extinct have been or are descended from a common ancestor or an ancestral gene pool. This last universal common ancestor of all organisms is believed to have appeared about 3.5 billion years ago. Biologists generally regard the universality of the genetic code as definitive evidence in favor of the theory of universal common descent (UCD) for all bacteria, archaea, and eukaryotes (see: origin of life).

Evolution does not always give rise to progressively more complex organisms. For example, the process of dysgenics has been observed among the human population.

Gene Theory

Biological form and function are created from and passed on to the next generation by genes, which are the primary units of inheritance. Physiological adaption to an organism's environment cannot be coded into its genes and cannot be inherited by its offspring (see Lamarckism). Remarkably, widely different organisms, including bacteria, plants, animals, and fungi, all share the same basic machinery that copies and transcribes DNA into proteins. For example, bacteria with inserted human DNA will correctly yield the corresponding human protein.

The total complement of genes in an organism or cell is known as its genome which is stored on one or more chromosomes. A chromosome is a single, long DNA strand on which thousands of genes, depending on the organism, are encoded. When a gene is active, the DNA code is transcribed into an RNA copy of the gene's information. A ribosome then translates the RNA into a structural protein or catalytic protein.

Homeostasis

Homeostasis is the ability of an open system to regulate its internal environment to maintain a stable condition by means of multiple dynamic equilibrium adjustments controlled by interrelated regulation mechanisms. All living organisms, whether unicellular or multicellular, exhibit homeostasis. Homeostasis exists at the cellular level, for example cells maintain a stable internal acidity (pH); and at the level of the organism, for example warm-blooded animals maintain a constant internal body temperature. Homeostasis is a term that is also used in association with ecosystems, for example, the atmospheric concentration of carbon dioxide on Earth has been regulated by the concentration of plant life on Earth because plants remove more carbon dioxide from the atmosphere during the daylight hours than they emit to the atmosphere at night. Tissues and organs can also maintain homeostasis.

Research

Structural

Molecular biology is the study of biology at a molecular level. This field overlaps with other areas of biology, particularly with genetics and biochemistry. Molecular biology chiefly concerns itself with understanding the interactions between the various systems of a cell, including the interrelationship of DNA, RNA, and protein synthesis and learning how these interactions are regulated.

Cell biology studies the physiological properties of cells, as well as their behaviors, interactions, and environment. This is done both on a microscopic and molecular level. Cell biology researches both single-celled organisms like bacteria and specialized cells in multicellular organisms like humans.

Understanding cell composition and how they function is fundamental to all of the biological sciences. Appreciating the similarities and differences between cell types is particularly important in the fields of cell and molecular biology. These fundamental similarities and differences provide a unifying theme, allowing the principles learned from studying one cell type to be extrapolated and generalized to other cell types.

Genetics is the science of genes, heredity, and the variation of organisms. Genes encode the information necessary for synthesizing proteins, which in turn play a large role in influencing (though, in many instances, not completely determining) the final phenotype of the organism. In modern research, genetics provides important tools in the investigation of the function of a particular gene, or the analysis of genetic interactions. Within organisms, genetic information generally is carried in chromosomes, where it is represented in the chemical structure of particular DNA molecules.

Developmental biology studies the process by which organisms grow and develop. Originating in embryology, modern developmental biology studies the genetic control of cell growth, differentiation, and "morphogenesis," which is the process that gives rise to tissues, organs, and anatomy. Model organisms for developmental biology include the round worm *Caenorhabditis elegans*, the fruit fly *Drosophila melanogaster*, the zebrafish *Brachydanio rerio*, the mouse *Mus musculus*, and the weed *Arabidopsis thaliana*.

Physiological

Physiology studies the mechanical, physical, and biochemical processes of living organisms by attempting to understand how all of the structures function as a whole. The theme of "structure to function" is central to biology. Physiological studies have traditionally been divided into plant physiology and animal physiology, but the principles of physiology are universal, no matter what particular organism is being studied. For example, what is learned about the physiology of yeast cells can also apply to human cells. The field of animal physiology extends the tools and methods of human physiology to non-human species. Plant physiology also borrows techniques from both fields.

Anatomy is an important branch of physiology and considers how organ systems in animals, such as the nervous, immune, endocrine, respiratory, and circulatory systems, function and interact. The study of these systems is shared with medically oriented disciplines such as neurology and immunology.

Evolution

Evolution is concerned with the origin and descent of species, as well as their change over time, and includes scientists from many taxonomically-oriented disciplines. For example, it generally involves scientists who have special training in particular organisms such as mammalogy, ornithology, botany, or herpetology, but use those organisms as systems to answer general questions about evolution. Evolutionary biology is mainly based on paleontology, which uses the fossil record to answer questions about the mode and tempo of evolution, as well as the developments in areas such as population genetics and evolutionary theory. In the 1980s, developmental biology re-entered evolutionary biology from its initial exclusion from the modern synthesis through the study of evolutionary developmental biology. Related fields which are often considered part of evolutionary biology are phylogenetics, systematics, and taxonomy.

Up into the 19th century, it was believed that life forms were being continuously created under certain conditions (see spontaneous generation). This misconception was challenged by William Harvey's diction that "all life [is] from [an] egg" (from the Latin "Omne vivum ex ovo"), a foundational concept of modern biology. It simply means that there is an unbroken continuity of life from its initial origin to the present time.

A group of organisms shares a common descent if they share a common ancestor. All organisms on the Earth have been and are descended from a common ancestor or an ancestral gene pool. This last universal common ancestor of all organisms is believed to have appeared about 3.5 billion years ago. Biologists generally regard the universality of the genetic code as definitive evidence in favor of the theory of universal common descent (UCD) for all bacteria, archaea, and eukaryotes (see: origin of life).

The two major traditional taxonomically-oriented disciplines are botany and zoology. Botany is the scientific study of plants. Botany covers a wide range of scientific disciplines that study the growth, reproduction, metabolism, development, diseases, and evolution of plant life. Zoology involves the study of animals, including the study of their physiology within the fields of anatomy and embryology. The common genetic and developmental mechanisms of animals and plants is studied in molecular biology, molecular genetics, and developmental biology. The ecology of animals is covered under behavioral ecology and other fields.

Taxonomy

Classification is the province of the disciplines of systematics and taxonomy. Taxonomy places organisms in groups called taxa, while systematics seeks to define their relationships with each other. This classification technique has evolved to reflect advances in cladistics and genetics, shifting the focus from physical similarities and shared characteristics to phylogenetics.

Traditionally, living things have been divided into five kingdoms:

Monera — Protista — Fungi — Plantae — Animalia

However, many scientists now consider this five-kingdom system to be outdated. Modern alternative classification systems generally begin with the three-domain system:

Archaea (originally Archaebacteria) — Bacteria (originally Eubacteria) — Eukarya

These domains reflect whether the cells have nuclei or not, as well as differences in the cell exteriors.

Further, each kingdom is broken down continuously until each species is separately classified. The order is:

1. Domain
2. Kingdom
3. Phylum

4. Class
5. Order
6. Family
7. Genus
8. Species

The scientific name of an organism is obtained from its genus and species. For example, humans would be listed as *Homo sapiens*. *Homo* would be the genus and *sapiens* is the species. Whenever writing the scientific name of an organism, it is proper to capitalize the first letter in the genus and put all of the species in lowercase; in addition the entire term would be put in italics or underlined. The term used for classification is called taxonomy.

There is also a series of intracellular parasites that are progressively "less alive" in terms of metabolic activity:

Viruses — Viroids — Prions

The dominant classification system is called Linnaean taxonomy, which includes ranks and binomial nomenclature. How organisms are named is governed by international agreements such as the International Code of Botanical Nomenclature (ICBN), the International Code of Zoological Nomenclature (ICZN), and the International Code of Nomenclature of Bacteria (ICNB). A fourth Draft BioCode was published in 1997 in an attempt to standardize naming in these three areas, but it has yet to be formally adopted. The Virus International Code of Virus Classification and Nomenclature (ICVCN) remains outside the BioCode.

Environmental

Ecology studies the distribution and abundance of living organisms, and the interactions between organisms and their environment. The environment of an organism includes both its habitat, which can be described as the sum of local abiotic factors such as climate and ecology, as well as the other the organisms that share its habitat. Ecological systems are studied at several different levels, from individuals and populations to ecosystems and the biosphere. As can be surmised, ecology is a science that draws on several disciplines.

Ethology studies animal behavior (particularly of social animals such as primates and canids), and is sometimes considered a branch of zoology. Ethologists have been particularly concerned with the evolution of behavior and the understanding of behavior in terms of the theory of natural selection. In one sense, the first modern ethologist was Charles Darwin, whose book "The Expression of the Emotions in Man and Animals" influenced many ethologists.

Biogeography studies the spatial distribution of organisms on the Earth, focusing on topics like plate tectonics, climate change, dispersal and migration, and cladistics.

Every living thing interacts with other organisms and its environment. One reason that biological systems can be difficult to study is that so many different interactions with other organisms and the environment are possible, even on the smallest of scales. A microscopic bacterium responding to a local sugar gradient is responding to its environment as much as a lion is responding to its environment when it searches for food in the African savannah. For any given species, behaviors can be co-operative, aggressive, parasitic or symbiotic. Matters become more complex when two or more different species interact in an ecosystem. Studies of this type are the province of ecology.

History

Although the concept of biology as a single coherent field arose in the 19th century, the biological sciences emerged from traditions of medicine and natural history reaching back to Galen and Aristotle in ancient Greece. During the Renaissance and early modern period, biological thought was revolutionized by a renewed interest in empiricism and the discovery of many novel organisms. Prominent in this movement were Vesalius and Harvey, who used experimentation and careful observation in physiology, and naturalists such as Linnaeus and Buffon who began to classify the diversity of life and the fossil record, as well as the development and behavior of organisms. Microscopy revealed the previously unknown world of microorganisms, laying the groundwork for cell theory. The growing importance of natural theology, partly a response to the rise of mechanical philosophy, encouraged the growth of natural history.

Over the 18th and 19th centuries, biological sciences such as botany and zoology became increasingly professional scientific disciplines. Lavoisier and other physical scientists began to connect the animate and inanimate worlds through physics and chemistry. Explorer-naturalists such as Alexander von Humboldt investigated the interaction between organisms and their environment, and the ways this relationship depends on geography—laying the foundations for biogeography, ecology and ethology. Naturalists began to reject essentialism and consider the importance of extinction and the mutability of species. Cell theory provided a new perspective on the fundamental basis of life. These developments, as well as the results from embryology and paleontology, were synthesized in Charles Darwin's theory of evolution by natural selection. The end of the 19th century saw the fall of spontaneous generation and the rise of the germ theory of disease, though the mechanism of inheritance remained a mystery.

In the early 20th century, the rediscovery of Mendel's work led to the rapid development of genetics by Thomas Hunt Morgan and his students, and by the 1930s the combination of population genetics and natural selection in the "neo-Darwinian synthesis". New disciplines developed rapidly, especially after Watson and Crick proposed the structure of DNA. Following the establishment of the Central Dogma and the cracking of the genetic code, biology was largely split between organismal biology—the fields that deal with whole organisms and groups of organisms—and the fields related to cellular and molecular biology. By the late 20th century, new fields like genomics and proteomics were reversing this trend, with organismal biologists using molecular techniques, and molecular and cell biologists investigating the interplay between genes and the environment, as well as the genetics of natural populations of organisms.

CELL THEORY

Cell theory refers to the idea that cells are the basic unit of structure of all living things. Development of this theory during the 1800s was made possible by advances in microscopy. This theory is one of the fundamental foundations of biology. The theory says that new cells are formed from other existing cells and the cell is a fundamental unit of structure, physiology, and organization in all living organisms.

History of Cell Theory

The cell was first discovered and named by Robert Hooke in 1665. He remarked that it looked strangely similar to *cellula* or small rooms which monks inhabited, thus deriving the name. However what Hooke actually saw was the dead cell walls of plant cells (cork) as it appeared under the microscope. Hooke's description of these cells was published in Micrographia. The cell walls observed by Hooke gave no indication of the nucleus and other organelles found in most living cells. The first man to witness a live cell under a microscope was Anton Van Leeuwenhoek, who in 1674 described the algae *Spirogyra*. Leeuwenhoek probably also saw bacteria.

The idea that cells were separable into individual units was proposed by Ludolph Christian Treviranus and Johann Jacob Paul Moldenhawer. All of this finally led to Henri Dutrochet formulating one of the fundamental tenets of modern cell theory by declaring that "The cell is the fundamental element of organization"

The observations of Hooke, Leeuwenhoek, Schleiden, Schwann, Virchow, and others led to the development of the cell theory. The cell theory is a widely accepted explanation of the relationship between cells and living things. The cell theory states:

- All living things are composed of cells.

- Cells are the basic unit of structure and function in living things.
- All cells are produced from other cells.

The cell theory holds true for all living things, no matter how big or small, or how simple or complex. Since cells are common to all living things, they can provide information about all life. And because all cells come from other cells, scientists can study cells to learn about growth, reproduction, and all other functions that living things perform. By learning about cells and how they function, you can learn about all types of living things.

Credit for developing Cell Theory is usually given to three scientists, Theodore Schwann, Matthias Jakob Schleiden, and Rudolf Virchow. In 1839 Schwann and Schleiden suggested that cells were the basic unit of life. Their theory accepted the first two tenets of modern cell theory (see next section, below). However the cell theory of Schleiden differed from modern cell theory in that it proposed a method of spontaneous crystallization that he called "Free Cell Formation". In 1858, Rudolf Virchow concluded that all cells come from pre-existing cells thus completing the classical cell theory.

Classical Cell Theory

1. All plants are made of cells (Schleiden)
2. All animals are made of cells (Schwann)
3. All cells come from pre-existing cells (Virchow)

Modern Cell Theory

The generally accepted parts of modern cell theory include:

1. The cell is the fundamental unit of structure and function in living things.
2. All cells come from pre-existing cells by division.
3. Energy flow (metabolism and biochemistry) occurs within cells.
4. Cells contain hereditary information which is passed from cell to cell during cell division
5. All cells are basically the same in chemical composition.
6. All known living things are made up of cells.
7. Some organisms are unicellular, made up of only one cell.
8. Others are multicellular, composed of countless number of cells.
9. The activity of an organism depends on the total activity of independent cells

Exceptions to the Theory

1. Viruses are considered by some to be alive, yet they are not made up of cells.
2. The first cell did not originate from a preexisting cell. See: Origin of life.

Types of Cells

Cells can be subdivided into the following subcategories: prokaryotes and eukaryotes. Prokaryotes lack a nucleus (though they do have circular DNA) and other membrane-bound organelles (though they do contain ribosomes). Eubacteria and Archeabacteria are two divisions of prokaryotes. Eukaryotes, on the other hand, have distinct nuclei and membrane-bound organelles (mitochondria, endoplasmic reticulum, vacuoles). In addition, they possess organized chromosomes which store genetic material.

RECOMBINANT DNA

Recombinant DNA is a form of artificial DNA which is engineered through the combination or insertion of one or more DNA strands, thereby combining DNA sequences which would not normally occur together. In terms of genetic modification, recombinant DNA is produced through the addition of relevant DNA into an existing organismal genome, such as the plasmid of bacteria, to code for or alter different traits for a specific purpose, such as immunity. It differs from genetic recombination, in that it does not occur through processes within the cell or ribosome, but is exclusively engineered.

The Recombinant DNA technique was engineered by Stanley Norman Cohen and Herbert Boyer in 1973. They published their findings in a 1974 paper entitled "Construction of Biologically Functional Bacterial Plasmids *in vitro*", which described a technique to isolate and amplify genes or DNA segments and insert them into another cell with precision, creating a transgenic bacterium. Recombinant DNA technology was made possible by the discovery of restriction endonucleases by Werner Arber, Daniel Nathans, and Hamilton Smith, for which they received the 1978 Nobel Prize in Medicine.

Introduction

Because of the importance of DNA in the replication of new structures and characteristics of living organisms, it has widespread importance in recapitulating via viral or non-viral vectors, both desirable and undesirable characteristics of a species to achieve characteristic change or to counteract effects caused by genetic or imposed disorders that have effects upon cellular or organismal processes. Through the use of recombinant DNA, genes that are identified as important can be amplified and isolated for use in other species or applications, where there may be some form of genetic illness or discrepancy, and provides a different approach to complex biological problem solving.

Applications and Methods

Cloning and Relation to Plasmids

The use of cloning is interrelated with Recombinant DNA in classical biology, as the term "clone" refers to a cell or organism derived from a parental organism, with modern biology referring to the term as a collection of cells derived from the same cell which remain identical. In the classical instance, the use of recombinant DNA provides the initial cell from which the host organism is then expected to recapitulate when it undergoes further cell division, with bacteria remaining a prime example due to the use of viral vectors in medicine which contain recombinant DNA inserted into a structure known as a plasmid.

Plasmids are extrachromosomal self replicating circular forms of DNA present in most bacteria, such as *Escherichia coli* (E. Coli), contain genes related to catabolism and metabolic activity, and allow the carrier bacterium to survive and reproduce in conditions present within other species and environments. These genes represent characteristics of resistance to bacteriophages and antibiotics and some heavy metals, but can also be fairly easily removed or separated from the plasmid by restriction endonucleases, which regularly produce "sticky ends" and allow the attachment of a selected segment of DNA, which codes for more "reparative" substances, such as peptide hormone medications including insulin, growth hormone, and oxytocin. In the introduction of useful genes into the plasmid, the bacteria is then used as a viral vector, which is encouraged to reproduce so as to recapitulate the altered DNA within other cells it infects and increase the amount of cells with the recombinant DNA present within them.

The use of plasmids is also key within gene therapy, where their related viruses are used as cloning vectors or carriers, which are means of transporting and passing on genes in recombinant DNA through viral reproduction throughout an organism. As a general definition of plasmids, the definition is that they contain three common features — a replicator, selectable marker and a cloning site. The replicator or "ori" refers to the origin of replication with regards to location and bacteria where replication begins. The marker refers to a gene which usually contains resistance to an antibiotic, but may also refer to a gene which is attached alongside the desired one, such as that which confers luminescence to allow identification of successfully recombined DNA. The cloning site is a sequence of nucleotides representing one or more positions where cleavage by restriction endonucleases occurs. Most eukaryotes do not maintain canonical plasmids; yeast is a notable exception. In addition, the Ti plasmid of the bacterium *Agrobacterium tumefaciens* can be used to integrate foreign DNA into the genomes of many plants. Other methods of introducing or creating recombinant DNA in eukaryotes include homologous recombination and transfection with modified viruses.

Chimeric Plasmids

When recombinant DNA is then further altered or changed to host additional strands of DNA, the molecule formed is referred to as "chimeric" DNA molecule, with reference to the mythological chimera which consisted as a composite of several animals. The presence of chimeric plasmid molecules is somewhat regular in occurrence as throughout the lifetime of an organism the propagation by vectors ensures the presence of hundreds of thousands of organismal and bacterial cells which all contain copies of the original chimeric DNA.

In the production of chimeric plasmids, the processes involved can be somewhat uncertain as the intended outcome of the addition of foreign DNA may not always be achieved and may result in the formation of unusable plasmids. Initially, the plasmid structure is linearised to allow the addition by bonding of complimentary foreign DNA strands to single-stranded "overhangs" or "sticky ends" present at the ends of the DNA molecule from staggered, or "S shaped" cleavages produced by restriction endonucleases.

A common vector used for the donation of plasmids originally was the bacterium Escherichia coli and later, the EcoRI derivative which was used for its versatility with addition of new DNA by "relaxed" replication when inhibited by chloramphenicol and spectinomycin; later being replaced by the pBR322 plasmid. In the case of EcoRI, the plasmid can anneal with the presence of foreign DNA via the route of sticky-end ligation, or with "blunt ends" via blunt-end ligation, in the presence of the phage T4 ligase, which forms covalent links between 3-carbon OH and 5-carbon PO4 groups present on blunt ends. Both sticky-end, or overhang ligation and blunt-end ligation can occur between foreign DNA segments, and cleaved ends of the original plasmid depending upon the restriction endonuclease used for cleavage.

Synthetic Insulin Production Using rDNA

In the 1920s there was no known way to produce insulin. The problem was solved, temporarily at least, when it was found that insulin from the pancreas of a pig could be used in humans. This method was the primary solution for diabetes until about six years ago. The problem with insulin production was raised again: there were not enough pigs to keep the supply of insulin intact. Scientists needed to devise another way. This is where one of the biggest breakthroughs in recombinant DNA technology happened.

The specific gene sequence, or oligonucleotide, that codes for insulin production in humans was introduced to a sample colony of *E. coli*. Only about 1 out of 106 bacteria picks up the sequence. This is not really a problem though, because the growth rate is only about 30 minutes for *E. coli*. In a 24 hour period there may be billions of *E. coli* that are coded with the DNA sequences needed to induce insulin production.

STEM CELL

Stem cells are cells found in all multi-cellular organisms. They retain the ability to renew themselves through mitotic cell division and can differentiate into a diverse range of specialized cell types. Research in the stem cell field grew out of findings by Canadian scientists Ernest A. McCulloch and James E. Till in the 1960s.

The two broad types of mammalian stem cells are: embryonic stem cells that are found in blastocysts, and adult stem cells that are are found in adult tissues. In a developing embryo, stem cells can differentiate into all of the specialized embryonic tissues. In adult organisms, stem cells and progenitor cells act as a repair system for the body, replenishing specialized cells, but also maintain the normal turnover of regenerative organs, such as blood, skin or intestinal tissues. There are also stem cells found in the amniotic fluid in the placenta, which is usually discarded after pregnancy.Also, stem cells can be found near the pelvic area, inside the anus. These stems cells usually smell really bad, obviously, these stem cells can only used to produce new buttockes, like for plastic surgery.

As stem cells can be grown and transformed into specialized cells with characteristics consistent with cells of various tissues such as muscles or nerves through cell culture, their use in medical therapies has been proposed. In particular, embryonic cell lines, autologous embryonic stem cells generated through therapeutic cloning, and highly plastic adult stem cells from the umbilical cord blood or bone marrow are touted as promising candidates.

Stem Cell Properties

Properties of Stem Cells

The classical definition of a stem cell requires that it possess two properties:

- *Self-renewal* - the ability to go through numerous cycles of cell division while maintaining the undifferentiated state.
- *Potency* - the capacity to differentiate into specialized cell types. In the strictest sense, this requires stem cells to be either totipotent or pluripotent - to be able to give rise to any mature cell type, although multipotent or unipotent progenitor cells are sometimes referred to as stem cells.

Potency Definitions

Potency specifies the differentiation potential (the potential to differentiate into different cell types) of the stem cell.

- Totipotent stem cells are produced from the fusion of an egg and sperm cell. Cells produced by the first few divisions of the fertilized egg are also totipotent. These cells can differentiate into embryonic and extraembryonic cell types.
- Pluripotent stem cells are the descendants of totipotent cells and can differentiate into cells derived from any of the three germ layers.
- Multipotent stem cells can produce only cells of a closely related family of cells (e.g. hematopoietic stem cells differentiate into red blood cells, white blood cells, platelets, etc.).
- Unipotent cells can produce only one cell type, but have the property of self-renewal which distinguishes them from non-stem cells (e.g. muscle stem cells).

Identifying Stem Cells

The practical definition of a stem cell is the functional definition - the ability to regenerate tissue over a lifetime. For example, the gold standard test for a bone marrow or hematopoietic stem cell (HSC) is the ability to transplant one cell and save an individual without HSCs. In this case, a stem cell must be able to produce new blood cells and immune cells over a long term, demonstrating potency. It should also be possible to isolate stem cells from the transplanted individual, which can themselves be transplanted into another individual without HSCs, demonstrating that the stem cell was able to self-renew.

Properties of stem cells can be illustrated *in vitro*, using methods such as clonogenic assays, where single cells are characterized by their ability to differentiate and self-renew. As well, stem cells can be isolated based on a distinctive set of cell surface markers. However, *in vitro* culture conditions can alter the behavior of cells, making it unclear whether the cells will behave in a similar manner *in vivo*. Considerable debate exists whether some proposed adult cell populations are truly stem cells.

Embryonic Stem Cells

Embryonic stem cell lines (ES cell lines) are cultures of cells derived from the epiblast tissue of the inner cell mass (ICM) of a blastocyst or earlier morula stage embryos. A blastocyst is an early stage embryo—approximately four to five days old in humans and consisting of 50–150 cells. ES cells are pluripotent and give rise during development to all derivatives of the three primary germ layers: ectoderm, endoderm and mesoderm. In other words, they can develop into each of the more than 200 cell types of the adult body when given sufficient and necessary stimulation for a specific cell type. They do not contribute to the extra-embryonic membranes or the placenta.

Nearly all research to date has taken place using mouse embryonic stem cells (mES) or human embryonic stem cells (hES). Both have the essential stem cell characteristics, yet they require very different environments in order to maintain an undifferentiated state. Mouse ES cells are grown on a layer of gelatin and require the presence of Leukemia Inhibitory Factor (LIF). Human ES cells are grown on a feeder layer of mouse embryonic fibroblasts (MEFs) and require the presence of basic Fibroblast Growth Factor (bFGF or FGF-2). Without optimal culture conditions or genetic manipulation, embryonic stem cells will rapidly differentiate.

A human embryonic stem cell is also defined by the presence of several transcription factors and cell surface proteins. The transcription factors Oct-4, Nanog, and Sox2 form the core regulatory network that ensures the suppression of genes that lead to differentiation and the maintenance of pluripotency. The cell surface antigens most commonly used to identify hES cells are the glycolipids SSEA3 and SSEA4 and the keratan sulfate antigens Tra-1-60 and Tra-1-81. The molecular definition of a stem cell includes many more proteins and continues to be a topic of research.

After twenty years of research, there are no approved treatments or human trials using embryonic stem cells. ES cells, being totipotent cells, require specific signals for correct differentiation - if injected directly into the body, ES cells will differentiate into many different types of cells, causing a teratoma. Differentiating ES cells into usable cells while avoiding transplant rejection are just a few of the hurdles that embryonic stem cell researchers still face. Many nations currently have moratoria on either ES cell research or the production of new ES cell lines. Because of their combined abilities of unlimited expansion and pluripotency, embryonic stem cells remain a theoretically potential source for regenerative medicine and tissue replacement after injury or disease.

Adult Stem Cells

The term adult stem cell refers to any cell which is found in a developed organism that has two properties: the ability to divide and create another cell like itself and also divide and create a cell more differentiated than itself. Also known as somatic (from Greek Óùìáôéêóò, "of the body") stem cells and germline (giving rise to gametes) stem cells, they can be found in children, as well as adults. Pluripotent adult stem cells are rare and generally small in number but can be found in a number of tissues including umbilical cord blood. Most adult stem cells are lineage-restricted (multipotent) and are generally referred to by their tissue origin (mesenchymal stem cell, adipose-derived stem cell, endothelial stem cell, etc.).

A great deal of adult stem cell research has focused on clarifying their capacity to divide or self-renew indefinitely and their differentiation potential. In mice, pluripotent stem cells are directly generated from adult fibroblast cultures.

While embryonic stem cell potential remains untested, adult stem cell treatments have been used for many years to treat successfully leukemia and related bone/blood cancers through bone marrow transplants. The use of adult stem cells in research and therapy is not as controversial as embryonic stem cells, because the production of adult stem cells does not require the destruction of an embryo. Consequently, more US government funding is being provided for adult stem cell research.

Lineage

To ensure self-renewal, stem cells undergo two types of cell division (see *Stem cell division and differentiation* diagram). Symmetric division gives rise to two identical daughter cells both endowed with stem cell properties. Asymmetric division, on the other hand, produces only one stem cell and a progenitor cell with limited self-renewal potential. Progenitors can go through several rounds of cell division before terminally differentiating into a mature cell. It is possible that the molecular distinction between symmetric and asymmetric divisions lies in differential segregation of cell membrane proteins (such as receptors) between the daughter cells.

An alternative theory is that stem cells remain undifferentiated due to environmental cues in their particular niche. Stem cells differentiate when they leave that niche or no longer receive those signals. Studies in *Drosophila* germarium have identified the signals dpp and adherins junctions that prevent germarium stem cells from differentiating.

The signals that lead to reprogramming of cells to an embryonic-like state are also being investigated. These signal pathways include several transcription factors including the oncogene c-Myc. Initial studies indicate that transformation of mice cells with a combination of these anti-differentiation signals can reverse differentiation and may allow adult cells to become pluripotent. However, the need to transform these cells with an oncogene may prevent the use of this approach in therapy.

Treatments

Medical researchers believe that stem cell therapy has the potential to change radically the treatment of human disease. A number of adult stem cell therapies already exist, particularly bone marrow transplants that are used to treat leukemia. In the future, medical researchers anticipate being able to use technologies derived from stem cell research to treat a wider variety of diseases including cancer, Parkinson's disease, spinal cord injuries, and muscle damage, amongst a number of other impairments and conditions. However, there still exists a great deal of social and scientific uncertainty surrounding stem cell research, which could possibly be overcome through public debate and future research.

Stem cells, however, are already used extensively in research, and some scientists do not see cell therapy as the first goal of the research, but see the investigation of stem cells as a goal worthy in itself.

Controversy Surrounding Human Embryonic Stem Cell Research

There exists a widespread controversy over human embryonic stem cell research that emanates from the techniques used in the creation and usage of stem cells. Human embryonic stem cell research is controversial because, with the present state of technology, starting a stem cell line requires the destruction of a human embryo and/or therapeutic cloning. However, recently, it has been shown in principle that embryonic stem cell lines can be generated using a single-cell biopsy similar to that used in preimplantation genetic diagnosis that may allow stem cell creation without embryonic destruction.

Adult stem cell research however carries none of the morality or ethics problems posed by human embryonic stem cell research. Further unlike embryonic research, the current adult stem cell research has been proven effective in treating many illnesses including but not limited to bone marrow transplants.

Recently Prof Ian Wilmut - The scientist who created Dolly the sheep, a breakthrough that provoked headlines around the world a decade ago - has decided to switch to adult stem cell research saying:

> "The odds are that by the time we make nuclear transfer work in humans, direct reprogramming will work too. I am anticipating that before too long we will be able to use the Yamanaka approach to achieve the same, without making human embryos. I have no doubt that in the long term, direct reprogramming will be more productive, though we can't be sure exactly when, next year or five years into the future."

Opponents of the research argue that embryonic stem cell technologies are a slippery slope to reproductive cloning and can fundamentally devalue human life. Those in the pro-life movement argue that a human embryo is a human life and is therefore entitled to protection.

Contrarily, supporters of embryonic stem cell research argue that such research should be pursued because the resultant treatments could have significant medical potential. It is also noted that excess embryos created for in vitro fertilisation could be donated with consent and used for the research.

The ensuing debate has prompted authorities around the world to seek regulatory frameworks and highlighted the fact that stem cell research represents a social and ethical challenge.

Key Stem Cell Research Events

- 1960s - Joseph Altman and Gopal Das present scientific evidence of adult neurogenesis, ongoing stem cell activity in the brain; their reports contradict Cajal's "no new neurons" dogma and are largely ignored.
- 1963 - McCulloch and Till illustrate the presence of self-renewing cells in mouse bone marrow.
- 1968 - Bone marrow transplant between two siblings successfully treats SCID.
- 1978 - Haematopoietic stem cells are discovered in human cord blood.
- 1981 - Mouse embryonic stem cells are derived from the inner cell mass by scientists Martin Evans, Matthew Kaufman, and Gail R. Martin. Gail Martin is attributed for coining the term "Embryonic Stem Cell".
- 1992 - Neural stem cells are cultured *in vitro* as neurospheres.
- 1997 - Leukemia is shown to originate from a haematopoietic stem cell, the first direct evidence for cancer stem cells.
- 1998 - James Thomson and coworkers derive the first human embryonic stem cell line at the University of Wisconsin-Madison.
- 2000s - Several reports of adult stem cell plasticity are published.
- 2001 - Scientists at Advanced Cell Technology clone first early (four- to six-cell stage) human embryos for the purpose of generating embryonic stem cells.
- 2003 - Dr. Songtao Shi of NIH discovers new source of adult stem cells in children's primary teeth.
- 2004-2005 - Korean researcher Hwang Woo-Suk claims to have created several human embryonic stem cell lines from unfertilised human oocytes. The lines were later shown to be fabricated.
- 2005 - Researchers at Kingston University in England claim to have discovered a third category of stem cell, dubbed cord-blood-derived embryonic-like stem

cells (CBEs), derived from umbilical cord blood. The group claims these cells are able to differentiate into more types of tissue than adult stem cells.

- August 2006 - the journal *Cell* publishes Kazutoshi Takahashi and Shinya Yamanaka, "Induction of Pluripotent Stem Cells from Mouse Embryonic and Adult Fibroblast Cultures by Defined Factors".
- October 2006 - Scientists in England create the first ever artificial liver cells using umbilical cord blood stem cells.
- January 2007 - Scientists at Wake Forest University led by Dr. Anthony Atala and Harvard University report discovery of a new type of stem cell in amniotic fluid. This may potentially provide an alternative to embryonic stem cells for use in research and therapy.
- June 2007 - Research reported by three different groups shows that normal skin cells can be reprogrammed to an embryonic state in mice. In the same month, scientist Shoukhrat Mitalipov reports the first successful creation of a primate stem cell line through somatic cell nuclear transfer
- October 2007 - Mario Capecchi, Martin Evans, and Oliver Smithies win the 2007 Nobel Prize for Physiology or Medicine for their work on embryonic stem cells from mice using gene targeting strategies producing genetically engineered mice (known as knockout mice) for gene research.
- November 2007 - Two similar papers released by their respective journals prior to formal publication: in *Cell* by Kazutoshi Takahashi and Shinya Yamanaka, "Induction of Pluripotent Stem Cells from Adult Human Fibroblasts by Defined Factors", and in *Science* by Junying Yu, et al., from the research group of James Thomson, "Induced Pluripotent Stem Cell Lines Derived from Human Somatic Cells": pluripotent stem cells generated from mature human fibroblasts. It is possible now to produce a stem cell from almost any other human cell instead of using embryos as needed previously, albeit the risk of tumorigenesis due to c-myc and retroviral gene transfer remains to be determined.

Stem Cell Funding & Policy Debate in the US

- 1995 - U.S. President Bill Clinton signs into law the Dickey Amendment which prohibited federally
- appropriated funds to be used for research where human embryos would be either created or destroyed.
- 02 November, 2004 - California voters approve Proposition 71, which provides $3 billion in state funds over ten years to human embryonic stem cell research.

- 2001-2006 - U.S. President George W. Bush endorses the Congress in providing federal funding for embryonic stem cell research of approximately $100 million as well as $250 million for research on adult and animal stem cells. He also enacts laws that restrict federally-funded stem cell research on embryonic stem cells to the already derived cell lines.
- 5 May, 2006 - Senator Rick Santorum introduces bill number S. 2754, or the Alternative Pluripotent Stem Cell Therapies Enhancement Act, into the U.S. Senate.
- 18 July, 2006 - The U.S. Senate passes the Stem Cell Research Enhancement Act H.R. 810 and votes down Senator Santorum's S. 2754.
- 19 July, 2006 - President George W. Bush vetoes H.R. 810 (Stem Cell Research Enhancement Act), a bill that would have reversed the Clinton-era law which made it illegal for federal money to be used for research where stem cells are derived from the destruction of an embryo.
- 07 November, 2006 - The people of the U.S. state of Missouri passed Amendment 2, which allows usage of any stem cell research and therapy allowed under federal law, but prohibits human reproductive cloning.
- 16 February, 2007 – The California Institute for Regenerative Medicine became the biggest financial backer of human embryonic stem cell research in the United States when they awarded nearly $45 million in research grants.

EVOLUTION

In biology, evolution is the change in the inherited traits of a population from one generation to the next. These traits are the expression of genes that are copied and passed on to offspring during reproduction. Mutations in these genes can produce new or altered traits, resulting in heritable differences between organisms. New traits can also come from transfer of genes between populations, as in migration, or between species, in horizontal gene transfer. Evolution occurs when these heritable differences become more common or rare in a population, either non-randomly through natural selection or randomly through genetic drift.

Natural selection is a process that causes *heritable* traits that are helpful for survival and reproduction to become more common, and harmful traits to become more rare. This occurs because organisms with advantageous traits pass on more copies of these heritable traits to the next generation. Over many generations, adaptations occur through a combination of successive, small, random changes in traits, and natural selection of those variants best-suited for their environment. In contrast, genetic drift produces random changes in the frequency of traits in a population. Genetic drift arises from the role chance plays in whether a given individual will survive and reproduce.

One definition of a species is a group of organisms that can reproduce with one another and produce fertile offspring. However, when a species is separated into populations that are prevented from interbreeding, mutations, genetic drift, and the selection of novel traits cause the accumulation of differences over generations and the emergence of new species. The similarities between organisms suggest that all known species are descended from a common ancestor (or ancestral gene pool) through this process of gradual divergence.

The theory of evolution by natural selection was proposed at about the same time by both Charles Darwin and Alfred Russel Wallace, and was set out in detail in Darwin's 1859 book *On the Origin of Species*. It encountered initial resistance from religious authorities who believed humans were divinely set apart from the animal kingdom. In the 1930s, Darwinian natural selection was combined with Mendelian inheritance to form the modern evolutionary synthesis, in which the connection between the *units* of evolution (genes) and the *mechanism* of evolution (natural selection) was made. This powerful explanatory and predictive theory has become the central organizing principle of modern biology, providing a unifying explanation for the diversity of life on Earth.

Heredity

Inheritance in organisms occurs through discrete traits – particular characteristics of an organism. In humans, for example, eye color is an inherited characteristic, which individuals can inherit from one of their parents. Inherited traits are controlled by genes and the complete set of genes within an organism's genome is called its genotype.

The complete set of observable traits that make up the structure and behavior of an organism is called its phenotype. These traits come from the interaction of its genotype with the environment. As a result, not every aspect of an organism's phenotype is inherited. Suntanned skin results from the interaction between a person's genotype and sunlight; thus, a suntan is not hereditary. However, people have different responses to sunlight, arising from differences in their genotype; a striking example is individuals with the inherited trait of albinism, who do not tan and are highly sensitive to sunburn.

Genes are regions within DNA molecules that contain genetic information. DNA is a long molecule with four types of bases attached along its length. Different genes have different sequences of bases; it is the sequence of these bases that encodes genetic information. Within cells, the long strands of DNA associate with proteins to form structures called chromosomes. A specific location within a chromosome is known as a locus. If the DNA sequence at a locus varies between individuals, the different forms of this sequence are called alleles. DNA sequences can change through mutations, producing new alleles. If a mutation occurs within a gene, the new allele may affect the trait that the gene controls, altering the phenotype of the organism. However, while this simple correspondence between an allele and a trait works in some cases, most traits are more complex and are controlled by multiple interacting genes.

Variation

Because an individual's phenotype results from the interaction of their genotype with the environment, the variation in phenotypes in a population reflects the variation in these organisms' genotypes. The modern evolutionary synthesis defines evolution as the change over time in this genetic variation. The frequency of one particular allele will fluctuate, becoming more or less prevalent relative to other forms of that gene. Evolutionary forces act by driving these changes in allele frequency in one direction or another. Variation disappears when an allele reaches the point of fixation — when it either disappears from the population or replaces the ancestral allele entirely.

Variation comes from mutations in genetic material, migration between populations (gene flow), and the reshuffling of genes through sexual reproduction. Variation also comes from exchanges of genes between different species; for example, through horizontal gene transfer in bacteria, and hybridization in plants. Despite the constant introduction of variation through these processes, most of the genome of a species is identical in all individuals of that species. However, even relatively small changes in genotype can lead to dramatic changes in phenotype: chimpanzees and humans differ in only about 5% of their genomes.

Mutation

Genetic variation comes from random mutations that occur in the genomes of organisms. Mutations are changes in the DNA sequence of a cell's genome and are caused by radiation, viruses, transposons and mutagenic chemicals, as well as errors that occur during meiosis or DNA replication. These mutagens produce several different types of change in DNA sequences; these can either have no effect, alter the product of a gene, or prevent the gene from functioning. Studies in the fly *Drosophila melanogaster* suggest that about 70 percent of mutations are deleterious, and the remainder are either neutral or have a weak beneficial effect. Due to the damaging effects that mutations can have on cells, organisms have evolved mechanisms such as DNA repair to remove mutations. Therefore, the optimal mutation rate for a species is a trade-off between short-term costs, such as the risk of cancer, and the long-term benefits of advantageous mutations.

Large sections of DNA can also be duplicated, which is a major source of raw material for evolving new genes, with tens to hundreds of genes duplicated in animal genomes every million years. Most genes belong to larger families of genes of shared ancestry. Novel genes are produced either through duplication and mutation of an ancestral gene, or by recombining parts of different genes to form new combinations with new functions. For example, the human eye uses four genes to make structures that sense light: three for color vision and one for night vision; all four arose from a single ancestral gene. An advantage of duplicating a gene (or even an (entire genome) is that overlapping or redundant functions in multiple genes allows alleles to be retained that would otherwise be harmful, thus increasing genetic diversity.

Changes in chromosome number may also involve the breakage and rearrangement of DNA within chromosomes. For example, two chromosomes in the *Homo* genus fused to produce human chromosome 2; this fusion did not occur in the lineage of the other apes, and they retain these separate chromosomes. The completion of both the human genome project and the chimpanzee genome project has allowed the identification of the relevant chromosomes. In evolution, the most important role of such chromosomal rearrangements may be to accelerate the divergence of a population into new species by preserving genetic differences within populations.

Sequences of DNA that can move about the genome, such as transposons, make up a major fraction of the genetic material of plants and animals, and may have been important in the evolution of genomes. For example, more than a million copies of the Alu sequence are present in the human genome, and these sequences have now been recruited to perform functions such as regulating gene expression. Another effect of these mobile DNA sequences is that when they move within a genome, they can mutate or delete existing genes and thereby produce genetic diversity.

Recombination

In asexual organisms, genes are inherited together, or *linked*, as they cannot mix with genes in other organisms during reproduction. However, the offspring of sexual organisms contain a random mixture of their parents' chromosomes that is produced through independent assortment. In the related process of genetic recombination, sexual organisms can also exchange DNA between two matching chromosomes. These shuffling processes can allow even alleles that are close together in a strand of DNA to be inherited independently. However, as only about one recombination event occurs per million base pairs in humans, genes close together on a chromosome may not be shuffled away from each other, and tend to be inherited together. This tendency is measured by finding how often two alleles occur together, which is called their linkage disequilibrium. A set of alleles that is usually inherited in a group is called a haplotype, and this co-inheritance can indicate that the locus is under positive selection.

Recombination in sexual organisms helps to remove harmful mutations and retain beneficial mutations. Consequently, when alleles cannot be separated by recombination – such as in mammalian Y chromosomes, which pass intact from fathers to sons – harmful mutations accumulate. In addition, recombination can produce individuals with new and advantageous gene combinations. These positive effects of recombination are balanced by the fact that this process can cause mutations and separate beneficial combinations of genes. The optimal rate of recombination for a species is therefore a trade-off between conflicting factors.

Mechanisms

There are three basic mechanisms of evolutionary change: natural selection, genetic drift, and gene flow. Natural selection favors genes that improve capacity for survival and reproduction. Genetic drift is random change in the frequency of alleles, caused by the random sampling of a generation's genes during reproduction, and gene flow is the transfer of genes within and between populations. The relative importance of natural selection and genetic drift in a population varies depending on the strength of the selection and the effective population size, which is the number of individuals capable of breeding. Natural selection usually predominates in large populations, while genetic drift dominates in small populations. The dominance of genetic drift in small populations can even lead to the fixation of slightly deleterious mutations. As a result, changing population size can dramatically influence the course of evolution. Population bottlenecks, where the population shrinks temporarily and therefore loses genetic variation, result in a more uniform population. Bottlenecks also result from alterations in gene flow such as decreased migration, expansions into new habitats, or population subdivision.

Natural Selection

Natural selection is the process by which genetic mutations that enhance reproduction become, and remain, more common in successive generations of a population. It has often been called a "self-evident" mechanism because it necessarily follows from three simple facts:

- Heritable variation exists within populations of organisms.
- Organisms produce more offspring than can survive.
- These offspring vary in their ability to survive and reproduce.

These conditions produce competition between organisms for survival and reproduction. Consequently, organisms with traits that give them an advantage over their competitors pass these advantageous traits on, while traits that do not confer an advantage are not passed on to the next generation.

The central concept of natural selection is the evolutionary fitness of an organism. This measures the organism's genetic contribution to the next generation. However, this is not the same as the total number of offspring: instead fitness measures the proportion of subsequent generations that carry an organism's genes. Consequently, if an allele increases fitness more than the other alleles of that gene, then with each generation this allele will become more common within the population. These traits are said to be "selected *for*". Examples of traits that can increase fitness are enhanced survival, and increased fecundity. Conversely, the lower fitness caused by having a less beneficial or deleterious allele results in this allele becoming rarer — they are "selected *against*". Importantly, the fitness of an allele is not a fixed characteristic, if the environment changes, previously neutral or harmful traits may become beneficial and previously beneficial traits become harmful.

Natural selection within a population for a trait that can vary across a range of values, such as height, can be categorized into three different types. The first is directional selection, which is a shift in the average value of a trait over time — for example organisms slowly getting taller. Secondly, disruptive selection is selection for extreme trait values and often results in two different values becoming most common, with selection against the average value. This would be when either short or tall organisms had an advantage, but not those of medium height. Finally, in stabilizing selection there is selection against extreme trait values on both ends, which causes a decrease in variance around the average value. This would, for example, cause organisms to slowly become all the same height.

A special case of natural selection is sexual selection, which is selection for any trait that increases mating success by increasing the attractiveness of an organism to potential mates. Traits that evolved through sexual selection are particularly prominent in males of some animal species, despite traits such as cumbersome antlers, mating calls or bright colors that attract predators, decreasing the survival of individual males. This survival disadvantage is balanced by higher reproductive success in males that show these hard to fake, sexually selected traits.

An active area of research is the unit of selection, with natural selection being proposed to work at the level of genes, cells, individual organisms, groups of organisms and even species. None of these models are mutually-exclusive and selection may act on multiple levels simultaneously. Below the level of the individual, genes called transposons try to copy themselves throughout the genome. Selection at a level above the individual, such as group selection, may allow the evolution of co-operation, as discussed below.

Genetic Drift

Genetic drift is the change in allele frequency from one generation to the next that occurs because alleles in the offspring generation are a random sample of those in the parent generation, and are thus subject to sampling error. As a result, when selective forces are absent or relatively weak, allele frequencies tend to "drift" upward or downward in a random walk. This drift halts when an allele eventually becomes fixed, either by disappearing from the population, or replacing the other alleles entirely. Genetic drift may therefore eliminate some alleles from a population due to chance alone, and two separate populations that began with the same genetic structure can drift apart by random fluctuation into two divergent populations with different sets of alleles. The time for an allele to become fixed by genetic drift depends on population size, with fixation occurring more rapidly in smaller populations.

Although natural selection is responsible for adaptation, the relative importance of the two forces of natural selection and genetic drift in driving evolutionary change in general is an area of current research in evolutionary biology. These investigations were prompted by the neutral theory of molecular evolution, which proposed that most evolutionary changes are the result the fixation of neutral mutations that do not have any immediate effects on the fitness of an organism. Hence, in this model, most genetic changes in a population are the result of constant mutation pressure and genetic drift.

Gene Flow

Gene flow is the exchange of genes between populations, which are usually of the same species. Examples of gene flow within a species include the migration and then breeding of organisms, or the exchange of pollen. Gene transfer between species includes the formation of hybrid organisms and horizontal gene transfer.

Migration into or out of a population can change allele frequencies..Immigration may add new genetic material to the established gene pool of a population. Conversely, emigration may remove genetic material. As barriers to reproduction between two diverging populations are required for the populations to become new species, gene flow may slow this process by spreading genetic differences between the populations. Gene flow is hindered by mountain ranges, oceans and deserts or even man-made structures such as the Great Wall of China, which has hindered the flow of plant genes.

Depending on how far two species have diverged since their most recent common ancestor, it may still be possible for them to produce offspring, as with horses and donkeys mating to produce mules. Such hybrids are generally infertile, due to the two different sets of chromosomes being unable to pair up during meiosis. In this case, closely-related species may regularly interbreed, but hybrids will be selected against and the species will remain distinct. However, viable hybrids are occasionally formed and these new species can either have properties intermediate between their parent species, or possess a totally new phenotype. The importance of hybridization in creating new species of animals is unclear, although cases have been seen in many types of animals, with the gray tree frog particularly well-studied.

Hybridization is, however, an important means of speciation in plants, since polyploidy (having more than two copies of each chromosome) is tolerated in plants more readily than in animals. Polyploidy is important in hybrids as it allows reproduction, with the two different sets of chromosomes each being able to pair with an identical partner during meiosis. Polyploids also have more genetic diversity, which allows them to avoid inbreeding depression in small populations.

Horizontal gene transfer is the transfer of genetic material from one organism to another organism that is not its offspring, this is most common among bacteria. In medicine, this contributes to the spread of antibiotic resistance, as when one bacteria acquires resistance genes it can rapidly transfer them to other species. Horizontal transfer of genes from bacteria to eukaryotes such as the yeast *Saccharomyces cerevisiae* and the adzuki bean beetle *Callosobruchus chinensis* may also have occurred. Viruses can also carry DNA between organisms, allowing transfer of genes even across biological domains. Gene transfer has also occurred within eukaryotic cells, from the chloroplast and mitochondrial genomes to nuclear genomes.

Outcomes

Evolution influences every aspect of the form and behavior of organisms. Most prominent are the specific behavioral and physical adaptations that are the outcome of natural selection. These adaptations increase fitness by aiding activities such as finding food, avoiding predators or attracting mates. Organisms can also respond to selection by co-operating with each other, usually by aiding their relatives or engaging in mutually-beneficial symbiosis. In the longer term, evolution produces new species through splitting ancestral populations of organisms into new groups that are unable to breed with one another.

These outcomes of evolution are sometimes divided into macroevolution, which is evolution that occurs at or above the level of species, such as speciation, and microevolution, which is smaller evolutionary changes, such as adaptations, within a species or population. In general, macroevolution is the outcome of long periods of microevolution. Thus, the distinction between micro- and macroevolution is not a fundamental one - the difference is simply the time involved. However, in macroevolution, the traits of the entire species are important. For instance, a large amount of variation among individuals allows a species to rapidly adapt to new habitats, lessening the chance of it going extinct, while a wide geographic range increases the chance of speciation, by making it more likely that part of the population will become isolated. In this sense, microevolution and macroevolution can sometimes be separate.

A common misconception is that evolution is "progressive," but natural selection has no long-term goal and does not necessarily produce greater complexity. Although complex species have evolved, this occurs as a side effect of the overall number of organisms increasing, and simple forms of life remain more common. For example, the overwhelming majority of species are microscopic prokaryotes, which form about half the world's biomass despite their small size, and constitute the vast majority of Earth's biodiversity. Simple organisms therefore remain the dominant form of life on Earth, and complex life appears more diverse only because it is more noticeable.

Adaptation

Adaptations are structures or behaviors that enhance a specific function, causing organisms to become better at surviving and reproducing. They are produced by a combination of the continuous production of small, random changes in traits, followed by natural selection of the variants best-suited for their environment. This process can cause either the gain of a new feature, or the loss of an ancestral feature. An example that shows both types of change is bacterial adaptation to antibiotic selection, with mutations causing antibiotic resistance by either modifying the target of the drug, or removing the transporters that allow the drug into the cell. However, many traits that appear to be simple adaptations are in fact exaptations: structures originally adapted for one function, but which coincidentally became somewhat useful for some other function in the process. One example is the African lizard *Holapsis guentheri*, which developed an extremely flat head for hiding in crevices, as can be seen by looking at its near relatives. However, in this species, the head has become so flattened that it assists in gliding from tree to tree - an exaptation.

As adaptation occurs through the gradual modification of existing structures, structures with similar internal organization may have very different functions in related organisms. This is the result of a single ancestral structure being adapted to function in different ways. The bones within bat wings, for example, are structurally similar to both human hands and seal flippers, due to the common descent of these structures from an ancestor that also had five digits at the end of each forelimb. Other idiosyncratic anatomical features, such as bones in the wrist of the panda being formed into a false "thumb," indicate that an organism's evolutionary lineage can limit what adaptations are possible. During adaption, some structures may lose their original function and become vestigial structures. Such structures may have little or no function in a current species, yet have a clear function in ancestral species, or other closely-related species. Examples include the non-functional remains of eyes in blind cave-dwelling fish, wings in flightless birds, and the presence of hip bones in whales and snakes. Examples of vestigial structures in humans include wisdom teeth, the coccyx, and the vermiform appendix.

An area of current investigation in evolutionary developmental biology is the developmental basis of adaptations and exaptations. This research addresses the origin and evolution of embryonic development and how modifications of development and developmental processes produce novel features. These studies have shown that evolution can alter development to create new structures, such as embryonic bone structures that develop into the jaw in other animals instead forming part of the middle ear in mammals. It is also possible for structures that have been lost in evolution to reappear due to changes in developmental genes, such as a mutation in chickens causing embryos to grow teeth similar to those of crocodiles.

Co-Evolution

Interactions between organisms can produce both conflict and co-operation. When the interaction is between pairs of species, such as a pathogen and a host, or a predator and its prey, these species can develop matched sets of adaptations. Here, the evolution of one species causes adaptations in a second species. These changes in the second species then, in turn, cause new adaptations in the first species. This cycle of selection and response is called co-evolution. An example is the production of tetrodotoxin in the rough-skinned newt and the evolution of tetrodotoxin resistance in its predator, the common garter snake. In this predator-prey pair, an evolutionary arms race has produced high levels of toxin in the newt and correspondingly high levels of resistance in the snake.

Co-Operation

However, not all interactions between species involve conflict. Many cases of mutually beneficial interactions have evolved. For instance, an extreme cooperation exists between plants and the mycorrhizal fungi that grow on their roots and aid the plant in absorbing nutrients from the soil. This is a reciprocal relationship as the plants provide the fungi with sugars from photosynthesis. Here, the fungi actually grow inside plant cells, allowing them to exchange nutrients with their hosts, while sending signals that suppress the plant immune system.

Coalitions between organisms of the same species have also evolved. An extreme case is the Eusociality found in social insects, such as bees, termites and ants, where sterile insects feed and guard the small number of organisms in a colony that are able to reproduce. On an even smaller scale, the somatic cells that make up the body of an animal are limited in their capacity to reproduce in order to maintain a stable organism which then supports a small number of the animal's germ cells to produce offspring. Here, somatic cells respond to specific signals that instruct them to either grow or kill themselves. If cells ignore these signals and attempt to multiply inappropriately, their uncontrolled growth causes cancer.

These examples of cooperation within species are thought to have evolved through the process of kin selection, which is where one organism acts to help raise a relative's offspring. This activity is selected for because if the *helping* individual contains alleles which promote the helping activity, it is likely that its kin will *also* contain these alleles and thus those alleles will be passed on. Other processes that may promote cooperation include group selection, where cooperation provides benefits to a group of organisms.

Speciation

Speciation is the process where a species diverges into two or more descendant species. It has been observed multiple times under both controlled laboratory conditions and in nature. In sexually-reproducing organisms, speciation results from reproductive isolation followed by genealogical divergence. There are four mechanisms for speciation. The most common in animals is allopatric speciation, which occurs in populations initially isolated geographically, such as by habitat fragmentation or migration. As selection and drift act independently in isolated populations, separation will eventually produce organisms that cannot interbreed.

The second mechanism of speciation is peripatric speciation, which occurs when small populations of organisms become isolated in a new environment. This differs from allopatric speciation in that the isolated populations are numerically much smaller than the parental population. Here, the founder effect causes rapid speciation through both rapid genetic drift and selection on a small gene pool.

The third mechanism of speciation is parapatric speciation. This is similar to peripatric speciation in that a small population enters a new habitat, but differs in that there is no physical separation between these two populations. Instead, speciation results from the evolution of mechanisms that reduce gene flow between the two populations. Generally this occurs when there has been a drastic change in the environment within the parental species' habitat. One example is the grass *Anthoxanthum odoratum*, which can undergo parapatric speciation in response to localized metal pollution from mines. Here, plants evolve that have resistance to high levels of metals in the soil. Selection against interbreeding with the metal-sensitive parental population produces a change in flowering time of the metal-resistant plants, causing reproductive isolation. Selection against hybrids between the two populations may cause *reinforcement*, which is the evolution of traits that promote mating within a species, as well as character displacement, which is when two species become more distinct in appearance.

Finally, in sympatric speciation species diverge without geographic isolation or changes in habitat. This form is rare since even a small amount of gene flow may remove genetic differences between parts of a population. Generally, sympatric speciation in animals requires the evolution of both genetic differences and non-random mating, to allow reproductive isolation to evolve.

One type of sympatric speciation involves cross-breeding of two related species to produce a new hybrid species. This is not common in animals as animal hybrids are usually sterile, because during meiosis the homologous chromosomes from each parent, being from different species cannot successfully pair. It is more common in plants, however because plants often double their number of chromosomes, to form polyploids. This allows the chromosomes from each parental species to form a matching pair during meiosis, as each parent's chromosomes is represented by a pair already. An example of such a speciation event is when the plant *Oenothera lamarckiana* gave rise to a new species of plant *O. gigas* in 1905, which had a different chromosome number than its parents. (O. gigas had a chromosome number of 2N = 28 while its parent *Oenothera lamarckiana* had a chromosome number of 2N = 14.)

Indeed, chromosome doubling can itself cause reproductive isolation, as half the doubled chromosomes will be unmatched when breeding with undoubled organisms.

Speciation events are important in the theory of punctuated equilibrium, which accounts for the pattern in the fossil record of short "bursts" of evolution interspersed with relatively long periods of stasis, where species remain relatively unchanged. In this theory, speciation and rapid evolution are linked, with natural selection and genetic drift acting most strongly on organisms undergoing speciation in novel habitats or small populations. As a result, the periods of stasis in the fossil record correspond to the parental population, and the organisms undergoing speciation and rapid evolution are found in small populations or geographically-restricted habitats, and therefore rarely being preserved as fossils.

Extinction

Extinction is the disappearance of an entire species. Extinction is not an unusual event, as species regularly appear through speciation, and disappear through extinction. Indeed, virtually all animal and plant species that have lived on earth are now extinct. These extinctions have happened continuously throughout the history of life, although the rate of extinction spikes in occasional mass extinction events. The Cretaceous–Tertiary extinction event, during which the dinosaurs went extinct, is the most well-known, but the earlier Permian-Triassic extinction event was even more severe, with approximately 96 percent of species driven to extinction. The Holocene extinction event is an ongoing mass extinction associated with humanity's expansion across the globe over the past few thousand years. Present-day extinction rates are 100-1000 times greater than the background rate, and up to 30 percent of species may be extinct by the mid 21st century. Human activities are now the primary cause of the ongoing extinction event; global warming may further accelerate it in the

The role of extinction in evolution depends on which type is considered. The causes of the continuous "low-level" extinction events, which form the majority of extinctions, are not well understood and may be the result of competition between species for shared resources. If competition from other species does alter the probability that a species will become extinct, this could produce species selection as a level of natural selection. The intermittent mass extinctions are also important, but instead of acting as a selective force, they drastically reduce diversity in a nonspecific manner and promote bursts of rapid evolution and speciation in survivors.

Evolutionary History of Life

Origin of Life

The origin of life is a necessary precursor for biological evolution, but understanding that evolution occurred once organisms appeared and investigating how this happens, does not depend on understanding exactly how life began. The current scientific consensus is that the complex biochemistry that makes up life came from simpler chemical reactions, but it is unclear how this occurred. Not much is certain about the earliest developments in life, the structure of the first living things, or the identity and nature of any last universal common ancestor or ancestral gene pool. Consequently, there is no scientific consensus on how life began, but proposals include self-replicating molecules such as RNA, and the assembly of simple cells.

Common Descent

All organisms on Earth are descended from a common ancestor or ancestral gene pool. Current species are a stage in the process of evolution, with their diversity the product of a long series of speciation and extinction events. The common descent of organisms was first deduced from four simple facts about organisms: First, they have geographic distributions that cannot be explained by local adaptation. Second, the diversity of life is not a set of completely unique organisms, but organisms that share morphological similarities. Third, vestigial traits with no clear purpose resemble functional ancestral traits, and finally, that organisms can be classified using these similarities into a hierarchy of nested groups.

Past species have also left records of their evolutionary history. Fossils, along with the comparative anatomy of present-day organisms, constitute the morphological, or anatomical, record. By comparing the anatomies of both modern and extinct species, paleontologists can infer the lineages of those species. However, this approach is most successful for organisms that had hard body parts, such as shells, bones or teeth. Further, as prokaryotes such as bacteria and archaea share a limited set of common morphologies, their fossils do not provide information on their ancestry.

More recently, evidence for common descent has come from the study of biochemical similarities between organisms. For example, all living cells use the same nucleic acids and amino acids. The development of molecular genetics has revealed the record of evolution left in organisms' genomes: dating when species diverged through the molecular clock produced by mutations. For example, these DNA sequence comparisons have revealed the close genetic similarity between humans and chimpanzees and shed light on when the common ancestor of these species existed.

Evolution of Life

Despite the uncertainty on how life began, it is clear that prokaryotes were the first organisms to inhabit Earth, approximately 3–4 billion years ago. No obvious changes in morphology or cellular organization occurred in these organisms over the next few billion years.

The eukaryotes were the next major innovation in evolution. These came from ancient bacteria being engulfed by the ancestors of eukaryotic cells, in a cooperative association called endosymbiosis. The engulfed bacteria and the host cell then underwent co-evolution, with the bacteria evolving into either mitochondria or hydrogenosomes. An independent second engulfment of cyanobacterial-like organisms led to the formation of chloroplasts in algae and plants.

The history of life was that of the unicellular eukaryotes, prokaryotes, and archaea until about a billion years ago when multicellular organisms began to appear in the oceans in the Ediacaran period. The evolution of multicellularity occurred in multiple independent events, in organisms as diverse as sponges, brown algae, cyanobacteria, slime moulds and myxobacteria.

Soon after the emergence of these first multicellular organisms, a remarkable amount of biological diversity appeared over approximately 10 million years, in an event called the Cambrian explosion. Here, the majority of types of modern animals evolved, as well as unique lineages that subsequently became extinct. Various triggers for the Cambrian explosion have been proposed, including the accumulation of oxygen in the atmosphere from photosynthesis. About 500 million years ago, plants and fungi colonized the land, and were soon followed by arthropods and other animals. Amphibians first appeared around 300 million years ago, followed by early amniotes, then mammals around 200 million years ago and birds around 100 million years ago (both from "reptile"-like lineages). However, despite the evolution of these large animals, smaller organisms similar to the types that evolved early in this process continue to be highly successful and dominate the Earth, with the majority of both biomass and species being prokaryotes.

History of Evolutionary Thought

Evolutionary ideas such as common descent and the transmutation of species have existed since at least the 6th century BC, when they were expounded by the Greek philosopher Anaximander. Evolutionary thought was further developed by other early thinkers, including the Greek philosopher Empedocles, the Roman philosopher Lucretius, the Arab biologist Al-Jahiz, the Persian philosopher Ibn Miskawayh, and the Brethren of Purity. Also in the Far East, the philosopher Zhuangzi discussed a transformative power of species to adapt to their surroundings. As biological knowledge grew in the 18th century, a variety of such ideas developed, beginning with Pierre Maupertuis in 1745, and with contributions from natural philosophers such as Erasmus Darwin and Jean-Baptiste Lamarck. In 1858, Charles Darwin and Alfred Russel Wallace jointly proposed the theory of evolution by natural selection to the Linnean Society of London in separate papers. Shortly after, Darwin's publication of *On the Origin of Species* provided detailed support for the theory and led to increasingly wide acceptance of the occurrence of evolution.

Nonetheless, Darwin's specific ideas about evolution, such as gradualism and the mechanisms of natural selection, were strongly contested at first. Lamarckists argued that transmutation of species occurred as parents passed on adaptations acquired during their lifetimes. Eventually, when experiments failed to support it, this idea was abandoned in favor of Darwinism. More significantly, Darwin could not account for how traits were passed down from generation to generation. A mechanism was provided in 1865 by Gregor Mendel, who found that traits were inherited in a predictable manner. When Mendel's work was rediscovered in 1900, disagreements over the rate of evolution predicted by early geneticists and biometricians led to a rift between the Mendelian and Darwinian models of evolution.

This contradiction was reconciled in the 1930s by biologists such as Ronald Fisher. The end result was a combination of evolution by natural selection and Mendelian inheritance, the modern evolutionary synthesis. In the 1940s, the identification of DNA as the genetic material by Oswald Avery and colleagues and the subsequent publication of the structure of DNA by James Watson and Francis Crick in 1953, demonstrated the physical basis for inheritance. Since then, genetics and molecular biology have become core parts of evolutionary biology and have revolutionized the field of phylogenetics.

In its early history, evolutionary biology primarily drew in scientists from traditional taxonomically-oriented disciplines, whose specialist training in particular organisms addressed general questions in evolution. As evolutionary biology expanded as an academic discipline, particularly after the development of the modern evolutionary synthesis, it began to draw more widely from the biological sciences. Currently the study of evolutionary biology involves scientists from fields as diverse as biochemistry, ecology, genetics and physiology, and evolutionary concepts are used in even more distant disciplines such as psychology, medicine, philosophy and computer science.

Social and Cultural Views

Even before the publication of *On the Origin of Species*, the idea that life had evolved was a source of debate and evolution is still a contentious concept. Debate has generally centered on the philosophical, social and religious implications of evolution, not on the science itself; the proposition that biological evolution occurs through the mechanism of natural selection is standard in the scientific literature.

As Darwin recognized early on, the most controversial aspect of evolutionary thought is its implications to humans. Specifically, some people believe that the origin of humans involved supernatural intervention, rather than natural processes. Although many religions and denominations have reconciled their beliefs with evolution through theistic evolution, several denominations contain creationists who object to evolution, as it contradicts their literal interpretation of origin beliefs. In some countries – notably the United States – these tensions between scientific and religious teachings have fueled the ongoing creation–evolution controversy, a religious conflict focusing on politics and public education. While other scientific fields such as cosmology and earth science also conflict with literal interpretations of many religious texts, evolutionary biology has strong oppositions from believers of religion.

Evolution has been used to support philosophical positions that promote discrimination and racism. For example, the eugenic ideas of Francis Galton were developed to argue that the human gene pool should be improved by selective breeding policies, including incentives for those considered "good stock" to reproduce, and the compulsory sterilization, prenatal testing, birth control, and even killing, of those considered *bad stock*. Another example of an extension of evolutionary theory that is now widely regarded as unwarranted is "Social Darwinism," a term given to the 19th century Whig Malthusian theory developed by Herbert Spencer into ideas about "survival of the fittest" in commerce and human societies as a whole, and by others into claims that social inequality, racism, and imperialism were justified. However, contemporary scientists and philosophers consider these ideas to have been neither mandated by evolutionary theory nor supported by data.

Applications in Technology

A major technological application of evolution is artificial selection, which is the intentional selection of certain traits in a population of organisms. Humans have used artificial selection for thousands of years in the domestication of plants and animals. More recently, such selection has become a vital part of genetic engineering, with selectable markers such as antibiotic resistance genes being used to manipulate DNA in molecular biology.

As evolution can produce highly optimized processes and networks, it has many applications in computer science. Here, simulations of evolution using evolutionary algorithms and artificial life started with the work of Nils Aall Barricelli in the 1960s, and was extended by Alex Fraser, who published a series of papers on simulation of artificial selection. Artificial evolution became a widely recognized optimization method as a result of the work of Ingo Rechenberg in the 1960s and early 1970s, who used evolution strategies to solve complex engineering problems. Genetic algorithms in particular became popular through the writing of John Holland. As academic interest grew, dramatic increases in the power of computers allowed practical applications. Evolutionary algorithms are now used to solve multi-dimensional problems more efficiently than software produced by human designers, and also to optimize the design of systems.

HOMEOSTASIS

Homeostasis is the property of either an open system or a closed system, especially a living organism, which regulates its internal environment so as to maintain a stable, constant condition. Multiple dynamic equilibrium adjustments and regulation mechanisms make homeostasis possible. The concept was created by Claude Bernard, often considered as the father of physiology, and published in 1865. The term was coined in 1932 by Walter Bradford Cannon from the Greek *homoios* (same, like, resembling) and *stasis* (to stand, posture).

Biological Homeostasis

With regard to any given life system parameter, an organism may be a *conformer* or a *regulator*. Regulators try to maintain the parameter at a constant level over possibly wide ambient environmental variations. On the other hand, conformers allow the environment to determine the parameter. For instance, endothermic animals maintain a constant body temperature, while ectothermic animals exhibit wide body temperature variation. Examples of endothermic animals include mammals and birds, examples of ectothermic animals include reptiles and some sea creatures.

This is not to say that conformers don't have behavioural adaptations allowing them to exert some control over a given parameter. For instance, reptiles often rest on sun-heated rocks in the morning to raise their body temperature. Likewise, regulators' behaviors may contribute to their internal stability: The same sun-baked rock may host a ground squirrel, also basking in the morning sun.

An advantage of homeostatic regulation is that it allows an organism to function effectively in a broad range of environmental conditions. For example, ectotherms tend to become sluggish at low temperatures, while a co-located endotherm may be fully active. That thermal stability comes at a price since an automatic regulation system requires additional energy. One reason snakes may eat only once a week is that they use much less energy to maintain homeostasis.

Most homeostatic regulation is controlled by the release of hormones into the bloodstream. However other regulatory processes rely on simple diffusion to maintain a balance.

Homeostatic regulation extends far beyond the control of temperature. All animals also regulate their blood glucose, as well as the concentration of their blood. Mammals regulate their blood glucose with insulin and glucagon. These hormones are released by the pancreas. If the pancreas is for any reason unable to produce enough of these two hormones diabetes results. The kidneys are used to remove excess water and ions from the blood. These are then expelled as urine. The kidneys perform a vital role in homeostatic regulation in mammals removing excess water, salt and urea from the blood. These are the body's main waste products.

Sleep timing depends upon a balance between homeostatic sleep propensity, the need for sleep as a function of the amount of time elapsed since the last adequate sleep episode, and circadian rhythms which determine the ideal timing of a correctly structured and restorative sleep episode.

Control Mechanisms

All homeostatic control mechanisms have at least three interdependent components for the variable being regulated: The receptor is the sensing component that monitors and responds to changes in the environment. When the receptor senses a stimulus, it sends information to a control center, the component that sets the range at which a variable is maintained. The control center determines an appropriate response to the stimulus. The result of that response feeds to the receptor, either enhancing it with positive feedback or depressing it with negative feedback.

Negative Feedback Mechanisms

Negative feedback mechanisms reduce or suppress the original stimulus, given the effector's output. Most homeostatic control mechanisms require a negative feedback loop to keep conditions from exceeding tolerable limits. The purpose is to prevent sudden severe changes within a complex organism. There are hundreds of negative feedback mechanisms in the human body. Among the most important regulatory functions are: thermoregulation, osmoregulation, and glucoregulation. The kidneys contribute to homeostasis in five important ways: regulation of blood water levels, reabsorption of substances into the blood, maintenance of salt and ion levels in the blood, regulation of blood pH, and excretion of urea and other wastes.

A negative feedback mechanism example is the typical home heating system. Its thermostat houses a thermometer, the receptor that senses when the temperature is too low. The control center, also housed in the thermostat, senses and responds to the thermometer when the temperature drops below a specified set point. Below that target level, the thermostat sends a message to the effector, the furnace. The furnace then produces heat, which warms the house. Once the thermostat senses a target level of heat has been reached, it will signal the furnace to turn off, thus maintaining a comfortable temperature - not too hot nor cold.

Positive Feedback Mechanisms

Positive feedback mechanisms are designed to accelerate or enhance the output created by a stimulus that has already been activated.

Unlike negative feedback mechanisms that initiate to maintain or regulate physiological functions within a set and narrow range, the positive feedback mechanisms are designed to push levels out of normal ranges. To achieve this purpose, a series of events initiates a cascading process that builds to increase the effect of the stimulus. This process can be beneficial but is rarely used by the body due to risks of the acceleration becoming uncontrollable.

One bodily positive feedback example event is blood platelet accumulation which in turn causes blood clotting in response to a break or tear in the lining of blood vessels. Another example is the release of oxytocin to intensify the contractions that take place during childbirth.

Positive feedback can also be harmful. An example being when you have a fever it causes a positive feedback within homeostasis that pushes the temperature continually higher. Body temperature can reach extremes of 45°C (113°F), at which cellular proteins denature, causing the active site in proteins to change, thus causing metabolism stop and ultimately resulting in death.

Homeostatic Imbalance

Much disease results from disturbance of homeostasis, a condition known as homeostatic imbalance. As it ages, every organism will lose efficiency in its control systems. The inefficiencies gradually result in an unstable internal environment that increases the risk for illness. In addition, homeostatic imbalance is also responsible for the physical changes associated with aging. Even more serious than illness and other characteristics of ageing, is death. Heart failure has been seen where nominal negative feedback mechanisms become overwhelmed, and destructive positive feedback mechanisms then take over.

Diseases which result from a homeostatic imbalance include diabetes, dehydration, hypoglycemia, hyperglycemia, gout and any disease caused by a toxin present in the bloodstream. All of these conditions result from the presence of an increased amount of a particular substance. Ideally homeostatic control mechanisms should prevent this imbalance from occurring but in some people the mechanisms do not work efficiently enough or the quantity of the substance exceeds the levels at which it can be managed. In these cases medical intervention is necessary to restore the imbalance or permanent damage to the organs may result.

Varieties of Homeostasis

The Dynamic Energy Budget theory for metabolic organisation delineates structure and (one or more) reserves in an organism. Its formulation is based on three forms of homeostasis:

- Strong homeostasis is where structure and reserve do not change in composition. Since the amount of reserve and structure can vary, this allows a particular change in the composition of the whole body (as explained by the Dynamic Energy Budget theory).
- Weak homeostasis is where the ratio of the amounts of reserve and structure becomes constant as long as food availability is constant, even when the organism grows. This means that the whole body composition is constant during growth in constant environments.
- Structural homeostasis means that the sub-individual structures grow in harmony with the whole individual; the relative proportions of the individuals remain constant.

Ecological Homeostasis

Ecological homeostasis is found in a climax community of maximum permitted biodiversity, given the prevailing ecological conditions.

In disturbed ecosystems or sub-climax biological communities such as the island of Krakatoa, after its major eruption in 1883, the established stable homeostasis of the previous forest climax ecosystem was destroyed and all life eliminated from the island. In the years after the eruption, Krakatoa went through a sequence of ecological changes in which successive groups of new plant or animal species followed one another, leading to increasing biodiversity and eventually culminating in a re-established climax community. This *ecological succession* on Krakatoa occurred in a number of several stages, in which a *sere* is defined as "a stage in a sequence of events by which succession occurs". The complete chain of seres leading to a climax is called a *prisere*. In the case of Krakatoa, the island as reached its climax community with eight hundred different species being recorded in 1983, one hundred years after the eruption which cleared all life off the island. Evidence confirms that this number has been homeostatic for some time, with the introduction of new species rapidly leading to elimination of old ones.

The evidence of Krakatoa, and other disturbed or virgin ecosystems shows that the initial colonisation by *pioneer* or *R strategy* species occurs through positive feedback reproduction strategies, where species are weeds, producing huge numbers of possible offspring, but investing little in the success of any one. Rapid boom and bust plague or pest cycles are observed with such species. As an ecosystem starts to approach climax these species get replaced by more sophisticated climax species which through negative feedback, adapt themselves to specific environmental conditions. These species, closely controlled by *carrying capacity*, follow *K strategies* where species produce fewer numbers of potential offspring, but invest more heavily in securing the reproductive success of each one to the micro-environmental conditions of its specific ecological niche.

It begins with a pioneer community and ends with a climax community. This climax community occurs when the ultimate vegetation has become in equilibrium with the local environment.

Such ecosystems form nested communities or *heterarchies*, in which homeostasis at one level, contributes to homeostatic processes at another holonic level. For example, the loss of leaves on a mature rainforest tree gives a space for new growth, and contributes to the plant litter and soil humus build-up upon which such growth depends. Equally a mature rainforest tree reduces the sunlight falling on the forest floor and helps prevent invasion by other species. But trees too fall to the forest floor and a healthy forest glade is dependent upon a constant rate of forest regrowth, produced by the fall of logs, and the recycling of forest nutrients through the respiration of termites and other insect, fungal and bacterial decomposers. Similarly such forest glades contribute ecological services, such as the regulation of microclimates or of the hydrological cycle for an ecosystem, and a number of different ecosystems act together to maintain homeostasis perhaps of a number of river catchments within a bioregion. A diversity of bioregions similarly makes up a stable homeostatic biological region or biome.

In the Gaia hypothesis, James Lovelock stated that the entire mass of living matter on Earth (or any planet with life) functions as a vast homeostatic superorganism that actively modifies its planetary environment to produce the environmental conditions necessary for its own survival. In this view, the entire planet maintains homeostasis. Whether this sort of system is present on Earth is still open to debate. However, some relatively simple homeostatic mechanisms are generally accepted. For example, when atmospheric carbon dioxide levels rise, certain plants are able to grow better and thus act to remove more carbon dioxide from the atmosphere. When sunlight is plentiful and atmospheric temperature climbs, the phytoplankton of the ocean surface waters thrive and produce more dimethyl sulfide, DMS. The DMS molecules act as cloud condensation nuclei which produce more clouds and thus increase the atmospheric albedo and this feeds back to lower the temperature of the atmosphere. As scientists discover more about Gaia, vast numbers of positive and negative feedback loops are being discovered, that together maintain a metastable condition, sometimes within very broad range of environmental conditions.

Reactive Homeostasis

Example of use: "Reactive homeostasis is an immediate response to a homeostatic challenge such as predation."

However, any homeostasis is impossible without reaction - because homeostasis is and must be a "feedback" phenomenon.

The phrase "reactive homeostasis" is simply short for: "reactive compensation reestablishing homeostasis", that is to say, "reestablishing a point of homeostasis." - it should not be confused with a separate *kind* of homeostasis or a distinct phenomenon *from* homeostasis, it is simply the compensation (or compensatory) phase of homeostasis.

Other Fields

The term has come to be used in other fields, as well.

Risk Homeostasis

An actuary may refer to *risk homeostasis*, where (for example) people who have anti-lock brakes have no better safety record than those without anti-lock brakes, because they unconsciously compensate for the safer vehicle via less-safe driving habits. Previously, certain maneuvers involved minor skids, evoking fear and avoidance: now the anti-lock system moves the boundary for such feedback, and behavior patterns expand into the no-longer punitive area. It has also been suggested that ecological crises are an instance of risk homeostasis in which behavior known to be dangerous continues until dramatic consequences actually occur.

Stress Homeostasis

Sociologists and psychologists may refer to *stress homeostasis*, the tendency of a population or an individual to stay at a certain level of stress, often generating artificial stresses if the "natural" level of stress is not enough.

Jean Francois Lyotard, a postmodern theorist, has applied this term to societal 'power centers' that he describes as being 'overned by a principle of homeostasis.' For example the scientific hierarchy, which will sometimes ignore a radical new discovery for years because it destabilizes previously accepted norms.

Waste Homeostasis

Andrew Potter has used the term *waste homeostasis* in reference to the lack of net gain from energy saving technologies.

Conversational Homeostasis

A 2007 study purported to find (and show clinically) *conversational homeostasis* in which overly-familiar people (such as spouses) condense their speech so much that they are actually worse at communicating novel information than strangers are, while not being conscious of this problem.

Metabolic Homeostasis

Some herbal medicines, known as adaptogens, have been defined to function as non-toxic metabolic regulators that can enhance metabolic homeostasis during stress.

CONSERVATION BIOLOGY

Conservation biology, or conservation ecology, is the science of analyzing and protecting Earth's biological diversity. Conservation biology draws from the biological, physical and social sciences, economics, and the practice of natural-resource management. Conservation ecology addresses population dynamics issues associated with the small population sizes of rare species (e.g., minimum viable populations). The term "conservation biology" refers to the application of science to the conservation of genes, populations, species, and ecosystems. Conservation biology is the scientific study of the phenomena that affect the maintenance, loss, and restoration of biological diversity. For the history of biodiversity conservation and volunteer activity, see conservation movement.

In the 19th century actions in the United Kingdom, the United States and certain other western countries emphasized the protection of habitat areas pursuant to visions of such people as John Muir and Theodore Roosevelt. It was not until the mid 20th century did efforts arise to target individual species for conservation, notably efforts in big cat conservation in South America led by the New York Zoological Society. By the 1970s, led primarily by work in the United States under the Endangered Species Act along with Biodiversity Action Plans developed in Australia, Sweden, the United Kingdom, hundreds of specific species protection plans ensued. The Society for Conservation Biology is a global community of conservation professionals dedicated to advancing the science and practice of conserving Earth's biological diversity.

Threats to Biological Diversity

Presently the Earth is undergoing the Holocene Mass Extinction, an era of unprecedented number of species extinctions. Human influence over the Earth's ecosystems has been so extensive within the last 10,000 years, that scientists have difficulty estimating the total number of species lost in this era; that is to say the rates of deforestation, reef destruction, wetlands filling and other human acts are proceeding much faster than human assessment of the Earth's species. The matter of ongoing species loss is made more complex by the fact that most of the Earth's species have not been described or evaluated for endangerment. The IUCN has found that 23 percent of vertebrates, 53 percent of invertebrates and 70 percent of plants that have been evaluated are designated as endangered or threatened. (The IUCN does not disaggregate endangered from critically endangered or threatened for the purpose of these statistics.)

Historically the main threat to biodiversity has been a set of threats generated from the overpopulation of humans: mass agriculture, deforestation, overgrazing, slash-and-burn, urban development, pesticide use. Worldwide, the effects of global warming add a potentially catastrophic threat to global biological diversity; a 2004 study by Chris Thomas, Lee Hannah, et al. estimated that 15 to 37 percent of all species would become extinct by 2050.

Importance of Biological Diversity

Biologist Bruce Walsh of the University of Arizona states three reasons for scientific interest in the preservation of species; genetic or medical resources, ecosystem stability, and ethics., and today the scientific community "stress[es] the importance" of maintaining biodiversity.

Biodiversity provides many ecosystem services that are often not readily visible. It plays an essential part in regulating the chemistry of our atmosphere, pollinating crops and generating water supply. Biodiversity is directly involved in recycling nutrients and providing fertile soils. Experiments with controlled environments have shown that humans cannot easily build ecosystems to support human needs; for example insect pollination cannot be mimicked by man-made construction. The total value of ecosystem services may amount to trillions of dollars in ecosystem services per annum to mankind. For example, one segment of North American forests has been assigned an annual value of 250 billion dollars; as another example, honey-bee pollination, a small segment of ecosystem services, is estimated to provide between 10 and 18 billion dollars of value per annum. The value of ecosystem services on one New Zealand island has been imputed to be as great as the GDP of that region.

History

Conservation biologists trace the ethics that guide their work back to early spiritual philosophies, including the Tao, Shinto, Hindu, Islamic and Buddhist traditions. In the West, origins of concern for the destruction of the natural environment by man can be traced to Plato; however, modern roots of conservation biology can be found in the late 18th century Enlightenment period particularly in England and Scotland. A number of thinkers, among them notably Lord Monboddo, described the importance of "preserving nature"; much of this early emphasis had its origins in Christian theology. By the early 1800s biogeography was ignited through efforts of Von Humboldt, DeCandolle, Lyell and Darwin; their efforts, while important in relating species to their environments, fell short of actual conservation.

The term *conservation* came into use in the late 19th century and referred to the management, mainly for economic reasons, of such natural resources as timber, fish, game, topsoil, pastureland, and minerals, and also to the preservation of forests (see forestry), wildlife (see wildlife refuge), parkland, wilderness, and watersheds. Western Europe was the source of much 19th century progress for conservation biology, particularly the British Empire; however, the United States began making sizable contributions to this field starting with thinking of Thoreau and taking form in the United States Congress passing the Forest Act of 1891, John Muir's work and the founding of the Sierra Club in 1895, founding of the New York Zoological Society in 1895 and establishment of a series of national forests and preserves by Theodore Roosevelt from 1901 to 1909.

In the early 20th century the New York Zoological Society was instrumental in developing concepts of establishing preserves for particular species and conducting the necessary conservation studies to determine the suitability of specific locations that are most appropriate as conservation priorities; the work of Henry Fairfield Osborn Jr., Archie Carr and Archie Carr III is notable in this era.

By the early 1970s national and international governmental agencies became more active in the conservation of biodiversity. Notably the United Nations acted to conserve sites of outstanding cultural or natural importance to the common heritage of mankind. The programme was adopted by the General Conference of UNESCO in 1972. As of 2006, a total of 830 sites are listed: 644 cultural, 162 natural. The first country to pursue aggressive biological conservation through national legislation was the USA, which passed back to back legislation in the Endangered Species Act (1966) and National Environmental Policy Act (1970), which together injected major funding and protection measures to large scale habitat protection and threatened species research.

By 1992 most of the countries of the world had become committed to the principles of conservation of biological diversity with the Convention on Biological Diversity; subsequently many countries began programmes of Biodiversity Action Plans to identify and conserve threatened species within their borders, as well as protect associated habitats.

The science of ecology has clarified the workings of the biosphere; i.e., the complex interrelationships among humans, other species, and the physical environment; moreover, the burgeoning human population, and associated agriculture, industry and its ensuing pollution have demonstrated how easily ecological relationships can be disrupted.

SYNTHETIC BIOLOGY

Synthetic biology has long been used to describe an approach to biology that attempts to integrate different areas of research in order to create a more holistic understanding of life. More recently the term has been used in a different way, signaling a new area of research that combines science and engineering in order to design and build novel biological functions and systems.

History

In 1978 the Nobel Prize in Physiology or Medicine was awarded to Werner Arber, Daniel Nathans and Hamilton O. Smith for the discovery of restriction enzymes and their application to problems of molecular genetics. In an editorial comment in the journal *Gene*, Wac³aw Szybalski wrote: "The work on restriction nucleases not only permits us easily to construct recombinant DNA molecules and to analyze individual genes but also has led us into *the new era of synthetic biology* where not only existing genes are described and analyzed but also new gene arrangements can be constructed and evaluated" (Gene 1978, 4, p 181).

Biology

Biologists are interested in learning more about how natural living systems work. One simple, direct way to test our current understanding of a natural living system is to build an instance (or version) of the system in accordance with our current understanding of the system. Michael Elowitz's early work on the Repressilator is one good example of such work. Elowitz had a model for how gene expression should work inside living cells. To test his model, he built a piece of DNA in accordance with his model, placed the DNA inside living cells, and watched what happened. Slight differences between observation and expectation highlight new science that may be well worth doing. Work of this sort often makes good use of mathematics to predict and study the dynamics of the biological system before experimentally constructing it. A wide variety of mathematical descriptions have been used with varying accuracy, including graph theory, Boolean networks, ordinary differential equations, stochastic differential equations, and Master equations (in order of increasing accuracy). Good examples include the work of Adam Arkin and Alexander van Oudenaarden. See also the PBS Nova special on artificial life.

Chemistry

Biological systems are physical systems that are made up of chemicals. Around 100 years ago, the science of chemistry went through a transition from studying natural chemicals to trying to design and build new chemicals. This transition led to the field of synthetic chemistry. In the same tradition, some aspects of synthetic biology can be viewed as an extension and application of synthetic chemistry to biology, and include work ranging from the creation of useful new biochemicals to studying the origins of life. Eric Kool's group at Stanford, Steven Benner's group at Florida, Carlos Bustamante's group at Berkeley, and Jack Szostak's group at Harvard are good examples of this tradition.

Engineering

Engineers view biology as a *technology*. Synthetic Biology includes the broad redefinition and expansion of biotechnology, with the ultimate goals of being able to design and build engineered biological systems that process information, manipulate chemicals, fabricate materials and structures, produce energy, provide food, and maintain and enhance human health and our environment. A good example of these technologies include the work of Chris Voigt, who redesigned the Type III secretion system used by *Salmonella typhimurium* to secrete spider silk proteins, a strong elastic biomaterial, instead of its own natural infectious proteins. One aspect of Synthetic Biology which distinguishes it from conventional genetic engineering is a heavy emphasis on developing foundational technologies that make the engineering of biology easier and more reliable. Good examples of engineering in Synthetic

Biology include the pioneering work of Tim Gardner and Jim Collins on an engineered genetic toggle switch, the Registry of Standard Biological Parts, and the International Genetically Engineered Machine competition (iGEM).

Re-Writing

Re-writers are Synthetic Biologists who are interested in testing the idea that since natural biological systems are so complicated, we would be better off re-building the natural systems that we care about, from the ground up, in order to provide engineered surrogates that are easier to understand and interact with. Re-writers draw inspiration from refactoring, a process sometimes used to improve computer software. Drew Endy and his group have done some preliminary work on re-writing (e.g., Refactoring Bacteriophage T7). Oligonucleotides harvested from a photolithographic or inkjet manufactured DNA chip combined with DNA mismatch error-correction allows inexpensive large-scale changes of codons in genetic systems to improve gene expression or incorporate novel amino-acids (see George Church's lab synthetic cell projects, Anthony Forster's lab and the Minimal Cell Project (MCP)). As in the T7 example above, this favors a synthesis-from-scratch approach.

Human Practices: Emerging Social, Ethical, Legal Challenges

In addition to numerous bioscientific challenges, the vast potential of synthetic biology to play a formative role in contemporary human life raises new questions for bioethics, biosecurity, biosafety, health, energy and intellectual property. To date considerable focus has been given to the so-called dual-use challenge. For example, while the study of synthetic biology can lead to more efficient ways to produce cures (e.g. against malaria), it may also lead to synthesis or redesign of harmful pathogens (e.g., smallpox). In addition, scientists, funders, policymakers, ethicists and others have recognized challenges presented by a post 9/11 political milieu. A new range of potentially malicious actors and actions (i.e., terrorists/terrorism) must now be taken into account by those seeking to govern scientific domains; and the internet and other new media provide global access to technological know-how and scientific knowledge. Such global access cannot be addressed using existing models of nation-specific regulation. (New Scientist, November 12, 2005). Some detailed suggestions for licensing and monitoring the various phases of gene and genome synthesis are beginning to appear. There is also an ongoing, comprehensive, and open discussion of so-called "societal issues" online at OpenWetWare.

Recently, efforts have been made to think beyond the "societal issues" model of ethics, politics, and science in relation to synthetic biology. This effort refuses the established convention of imagining society outside of and downstream of scientific practices, such that bioethics is assigned the task of limiting the negative impact of science on society. By contrast recent approaches focus on the integral and mutually formative relations among scientific and other human practices. These human practices approaches attempt to invent ongoing and regular forms of collaboration among synthetic biologists, ethicists, political analysts, funders, human scientists and civil society activists. To date collaborative work on "governance" or "society" or "ethics" in relation to synthetic biology has largely consisted either of intensive, short term meetings, aimed at producing guidelines or regulations, or standing committees whose purpose is limited to protocol review or rule enforcement. Such work has proven valuable in identifying the ways in which synthetic biology intensifies already-known challenges in rDNA technologies. However, these forms are not suited to identifying new challenges as they emerge. An example of efforts to develop ongoing collaboration is the Human Practices component of the Synthetic Biology Engineering Research Center (SynBERC), an NSF funded collaboration among a number of leading research universities. In Europe, the multi-partner project SYNBIOSAFE is investigating the biosafety, biosecurity and ethical aspects of synthetic biology.

Key Enabling Technologies

There are several key enabling technologies that are critical to the growth of synthetic biology.

Sequencing

Synthetic biologists make use of DNA sequencing in their work in several ways. First, large-scale genome sequencing efforts continue to provide a wealth of information on naturally occurring organisms. This information provides a rich substrate from which synthetic biologists can construct parts and devices. Second, synthetic biologists use sequencing to verify that they fabricated their engineered system as intended. Third, fast, cheap and reliable sequencing can also facilitate rapid detection and identification of synthetic systems and organisms.

Fabrication

A critical limitation in synthetic biology today is the time and effort expended during fabrication of engineered biological systems. As of spring 2007, synthesis of a 1000bp gene costs approximately $800 and 2 weeks to construct. To speed up the cycle of design, fabrication, testing and redesign, synthetic biology requires more rapid and reliable *de novo* DNA synthesis and assembly of fragments of DNA.

Modeling

Models inform the design of engineered biological systems by allowing synthetic biologists to better predict system behavior prior to fabrication. Synthetic biology will benefit from better models of how biological molecules bind substrates and catalyze reactions, how DNA encodes the information needed to specify the cell and how multi-component integrated systems behave.

Measurement

Precise and accurate quantitative measurements of biological systems are crucial to improving understanding of biology. Such measurements often help to elucidate how biological systems work and provide the basis for model construction and validation. Differences between predicted and measured system behavior can identify gaps in understanding and explain why synthetic systems don't always behave as intended. Technologies which allow many parallel and time-dependent measurements will be especially useful in synthetic biology. Microscopy and flow cytometry are examples of useful measurement technologies.

REFERENCES

Allen, GE (1978). Life Science in the Twentieth Century. Cambridge University Press.

Antony van Leeuwenhoek: tercentenary of his discovery of bacteria by J. R. Porter in *Bacteriol. Rev.* (1976) Volume 40, pages 260–269.

Coleman, W (1978). Biology in the Nineteenth Century: Problems of Form, Function and Transformation. Cambridge University Press.

Darwin, Charles (1859). On the Origin of Species, 1st, John Murray

Dutrochet, Henri 1824, "Anatomical and Physiological Researches on the Intimate Structures of Animals and Plants, and Their Motility."

Fruton, JS (1999). Proteins, Enzymes, Genes: The Interplay of Chemistry and Biology. Yale University Press.

Futuyma, DJ (2005). Evolution. Sinauer Associates.

King, TJ & Roberts, MBV (1986). Biology: A Functional Approach. Thomas Nelson and Sons.

Lynn, Richard; Van Court, Marilyn (2004). "New evidence of dysgenic fertility for intelligence in the United States". Intelligence 32 (2): p. 193. Ablex Pub.

Magner, LN (2002). A History of the Life Sciences. TF-CRC.

Margulis, L; Schwartz, KV (1997). Five Kingdoms: An Illustrated Guide to the Phyla of Life on Earth, 3rd edition, WH Freeman & Co.

Mayr, E (1985). The Growth of Biological Thought. Belknap Press.

Mazzarello, P (1999). "A unifying concept: the history of cell theory". Nature Cell Biology 1: E13-E15.

Moldenhawer, Johann Jacob Paul 1812, "Beyträge zur Anatomie der Pflanzen."

Morange, M & Cobb, M (2000). A History of Molecular Biology. Harvard University Press.

Schleiden, Matthias Jakob 1839, "Contributions to Phytogenesis."

Smocovitis, VB (1996). Unifying Biology. Princeton University Press.

The American Naturalist, Vol.73 pgs 517-537.

Treviranus, Ludolph Christian 1811, "Beyträge zur Pflanzenphysiologie."

Woese C, Kandler O, Wheelis M (1990). "Towards a natural system of organisms: proposal for the domains Archaea, Bacteria, and Eukarya.". Proc Natl Acad Sci U S A 87 (12): 4576-9.

2

Introductory Overview of Ecology and Environmental Science

ECOLOGY

Ecology is the scientific study of the distribution and abundance of living organisms and the interactions among organisms and their environment. The environment of an organism includes both physical properties, which can be described as the sum of local abiotic factors such as insolation (sunlight), climate, and geology, and biotic factors, which are other organisms that share its habitat.

The word "ecology" is often used more loosely in such terms as social ecology and deep ecology and in common parlance as a synonym for the natural environment or environmentalism. Likewise "ecologic" or "ecological" is often taken in the sense of environmentally friendly.

The term ecology or *oekologie* was coined by the German biologist Ernst Haeckel in 1866, when he defined it as "the comprehensive science of the relationship of the organism to the environment."

Haeckel did not elaborate on the concept, and the first significant textbook on the subject (together with the first university course) was written by the Danish botanist, Eugenius Warming. For this early work, Warming is often identified as the founder of ecology.

Scope

Ecology is usually considered a branch of biology, the general science that studies living organisms. Organisms can be studied at many different levels, from proteins and nucleic acids (in biochemistry and molecular biology), to cells (in cellular biology), to individuals (in botany, zoology, and other similar disciplines), and finally at the level of populations, communities, and ecosystems, to the biosphere as a whole; these latter strata are the primary subjects of ecological inquiry. Ecology is a multi-disciplinary science. Because of its focus on the higher levels of the organization of life on earth and on the interrelations between organisms and their environment, ecology draws heavily on many other branches of science, especially geology and geography, meteorology, pedology, genetics, chemistry, and physics. Thus, ecology is considered by some to be a holistic science, one that over-arches older disciplines such as biology which in this view become sub-disciplines contributing to ecological knowledge. In support of viewing ecology as a subject in its own right as opposed to a sub-discipline of biology, Robert Ulanowicz stated that "*The emerging picture of ecosystem behavior does not resemble the worldview imparted by an extrapolation of conceptual trends established in other sciences.*"

Agriculture, fisheries, forestry, medicine and urban development are among human activities that would fall within Krebs' (1972: 4) explanation of his definition of ecology: *where organisms are found, how many occur there, and why.*

Ecological knowledge such as the quantification of biodiversity and population dynamics have provided a scientific basis for expressing the aims of environmentalism and evaluating its goals and policies. Additionally, a holistic view of nature is stressed in both ecology and environmentalism.

Consider the ways an ecologist might approach studying the life of honeybees:

- The behavioral relationship between individuals of a species is behavioral ecology — for example, the study of the queen bee, and how she relates to the worker bees and the drones.
- The organized activity of a species is community ecology; for example, the activity of bees assures the pollination of flowering plants. Bee hives additionally produce honey which is consumed by still other species, such as bears.
- The relationship between the environment and a species is environmental ecology — for example, the consequences of environmental change on bee activity. Bees may die out due to environmental changes (see pollinator decline). The environment simultaneously affects and is a consequence of this activity and is thus intertwined with the survival of the species.

Disciplines of Ecology

Ecology is a broad discipline comprising many sub-disciplines. A common, broad classification, moving from lowest to highest complexity, where complexity is defined as the number of entities and processes in the system under study, is:

- Ecophysiology and Behavioral ecology examine adaptations of the individual to its environment.
- Population ecology studies the dynamics of populations of a single species.
- Community ecology (or synecology) focuses on the interactions between species within an ecological community.
- Ecosystem ecology studies the flows of energy and matter through the biotic and abiotic components of ecosystems.
- Systems ecology is an interdisciplinary field focusing on the study, development, and organization of ecological systems from a holistic perspective.
- Landscape ecology examines processes and relationship across multiple ecosystems or very large geographic areas.

Ecology can also be sub-divided according to the species of interest into fields such as animal ecology, plant ecology, insect ecology, and so on. Another frequent method of subdivision is by biome studied, e.g., Arctic ecology (or polar ecology), tropical ecology, desert ecology, etc. The primary technique used for investigation is often used to subdivide the discipline into groups such as chemical ecology, genetic ecology, field ecology, statistical ecology, theoretical ecology, and so forth. These fields are not mutually exclusive; one could be a theoretical plant community ecologist, or a polar ecologist interested in animal genetics.

History of Ecology: Fundamental Principles of Ecology

Biosphere

For modern ecologists, ecology can be studied at several levels: population level (individuals of the same species in the same or similar environment), biocoenosis level (or community of species), ecosystem level, and biosphere level.

The outer layer of the planet Earth can be divided into several compartments: the hydrosphere (or sphere of water), the lithosphere (or sphere of soils and rocks), and the atmosphere (or sphere of the air). The biosphere (or sphere of life), sometimes described as "the fourth envelope", is all living matter on the planet or that portion of the planet occupied by life. It reaches well into the other three spheres, although there are no permanent inhabitants of the atmosphere. Relative to the volume of the Earth, the biosphere is only the very thin surface layer which extends from 11,000 meters below sea level to 15,000 meters above.

It is thought that life first developed in the hydrosphere, at shallow depths, in the photic zone. (Recently, though, a competing theory has emerged, that life originated around hydrothermal vents in the deeper ocean. See Origin of life.) Multicellular organisms then appeared and colonized benthic zones. Photosynthetic organisms gradually produced the chemically unstable oxygen-rich atmosphere that characterizes our planet. Terrestrial life developed later, after the ozone layer protecting living beings from UV rays formed. Diversification of terrestrial species is thought to be increased by the continents drifting apart, or alternately, colliding. Biodiversity is expressed at the ecological level (ecosystem), population level (intraspecific diversity), species level (specific diversity), and genetic level. Recently technology has allowed the discovery of the deep ocean vent communities. This remarkable ecological system is not dependent on sunlight but bacteria, utilising the chemistry of the hot volcanic vents, are at the base of its food chain.

The biosphere contains great quantities of elements such as carbon, nitrogen, hydrogen and oxygen. Other elements, such as phosphorus, calcium, and potassium, are also essential to life, yet are present in smaller amounts. At the ecosystem and biosphere levels, there is a continual recycling of all these elements, which alternate between the mineral and organic states.

While there is a slight input of geothermal energy, the bulk of the functioning of the ecosystem is based on the input of solar energy. Plants and photosynthetic microorganisms convert light into chemical energy by the process of photosynthesis, which creates glucose (a simple sugar) and releases free oxygen. Glucose thus becomes the secondary energy source which drives the ecosystem. Some of this glucose is used directly by other organisms for energy. Other sugar molecules can be converted to other molecules such as amino acids. Plants use some of this sugar, concentrated in nectar to entice pollinators to aid them in reproduction.

Cellular respiration is the process by which organisms (like mammals) break the glucose back down into its constituents, water and carbon dioxide, thus regaining the stored energy the sun originally gave to the plants. The proportion of photosynthetic activity of plants and other photosynthesizers to the respiration of other organisms determines the specific composition of the Earth's atmosphere, particularly its oxygen level. Global air currents mix the atmosphere and maintain nearly the same balance of elements in areas of intense biological activity and areas of slight biological activity.

Water is also exchanged between the hydrosphere, lithosphere, atmosphere and biosphere in regular cycles. The oceans are large tanks, which store water, ensure thermal and climatic stability, as well as the transport of chemical elements thanks to large oceanic currents.

For a better understanding of how the biosphere works, and various dysfunctions related to human activity, American scientists simulated the biosphere in a small-scale model, called Biosphere II.

The Ecosystem Concept

The first principle of ecology is that each living organism has an ongoing and continual relationship with every other element that makes up its environment. An ecosystem can be defined as any situation where there is interaction between organisms and their environment.

The ecosystem is of two entities, the entirety of life, the biocoenosis, and the medium that life exists in, the biotope. Within the ecosystem, species are connected by food chains or food webs. Energy from the sun, captured by primary producers via photosynthesis, flows upward through the chain to primary consumers (herbivores), and then to secondary and tertiary consumers (carnivores and omnivores), before ultimately being lost to the system as waste heat. In the process, matter is incorporated into living organisms, which return their nutrients to the system via decomposition, forming biogeochemical cycles such as the carbon and nitrogen cycles.

The concept of an ecosystem can apply to units of variable size, such as a pond, a field, or a piece of dead wood. An ecosystem within another ecosystem is called a micro ecosystem. *For example, an ecosystem can be a stone and all the life under it.* A meso ecosystem *could be a forest, and a* macro ecosystem *a whole eco region, with its drainage basin.*

The main questions when studying an ecosystem are:

- Whether the colonization of a barren area could be carried out
- Investigation the ecosystem's dynamics and changes
- The methods of which an ecosystem interacts at local, regional and global scale
- Whether the current state is stable
- Investigating the value of an ecosystem and the ways and means that interaction of ecological systems provides benefits to humans, especially in the provision of healthy water.

Ecosystems are often classified by reference to the biotopes concerned. The following ecosystems may be defined:

- As continental ecosystems, such as forest ecosystems, meadow ecosystems such as steppes or savannas, or agro-ecosystems
- As ecosystems of inland waters, such as lentic ecosystems such as lakes or ponds; or lotic ecosystems such as rivers
- As oceanic ecosystems.

Another classification can be done by reference to its communities, such as in the case of an human ecosystem.

Dynamics and Stability

Ecological factors which affect dynamic change in a population or species in a given ecology or environment are usually divided into two groups: abiotic and biotic.

Abiotic factors are geological, geographical, hydrological and climatological parameters. A biotope is an environmentally uniform region characterized by a particular set of abiotic ecological factors. Specific abiotic factors include:

- Water, which is at the same time an essential element to life and a milieu
- Air, which provides oxygen, nitrogen, and carbon dioxide to living species and allows the dissemination of pollen and spores
- Soil, at the same time source of nutriment and physical support
    - Soil pH, salinity, nitrogen and phosphorus content, ability to retain water, and density are all influential
- Temperature, which should not exceed certain extremes, even if tolerance to heat is significant for some species
- Light, which provides energy to the ecosystem through photosynthesis
- Natural disasters can also be considered abiotic

Biocenose, or community, is a group of populations of plants, animals, micro-organisms. Each population is the result of procreations between individuals of same species and cohabitation in a given place and for a given time. When a population consists of an insufficient number of individuals, that population is threatened with extinction; the extinction of a species can approach when all biocenoses composed of individuals of the species are in decline. In small populations, consanguinity (inbreeding) can result in reduced genetic diversity that can further weaken the biocenose.

Biotic ecological factors also influence biocenose viability; these factors are considered as either intraspecific and interspecific relations.

Intraspecific relations are those which are established between individuals of the same species, forming a population. They are relations of co-operation or competition, with division of the territory, and sometimes organization in hierarchical societies.

Interspecific relations—interactions between different species—are numerous, and usually described according to their beneficial, detrimental or neutral effect (for example, mutualism (relation ++) or competition (relation —). The most significant relation is the relation of predation (to eat or to be eaten), which leads to the essential concepts in ecology of food chains (for example, the grass is consumed by the herbivore, itself consumed by a carnivore, itself consumed by a carnivore of larger size). A high predator to prey ratio can have a negative influence on both the predator and prey biocenoses in that low availability of food and high death rate prior to sexual maturity can decrease (or prevent the increase of) populations of each, respectively. Selective hunting of species by humans which leads to population decline is one example of a high predator to prey ratio in action. Other interspecific relations include parasitism, infectious disease and competition for limiting resources, which can occur when two species share the same ecological niche.

The existing interactions between the various living beings go along with a permanent mixing of mineral and organic substances, absorbed by organisms for their growth, their maintenance and their reproduction, to be finally rejected as waste. These permanent recyclings of the elements (in particular carbon, oxygen and nitrogen) as well as the water are called biogeochemical cycles. They guarantee a durable stability of the biosphere (at least when unchecked human influence and extreme weather or geological phenomena are left aside). This self-regulation, supported by negative feedback controls, ensures the perenniality of the ecosystems. It is shown by the very stable concentrations of most elements of each compartment. This is referred to as homeostasis. The ecosystem also tends to evolve to a state of ideal balance, reached after a succession of events, the climax (for example a pond can become a peat bog).

Spatial Relationships and Subdivisions of Land

Ecosystems are not isolated from each other, but are interrelated. For example, water may circulate between ecosystems by the means of a river or ocean current. Water itself, as a liquid medium, even defines ecosystems. Some species, such as salmon or freshwater eels move between marine systems and fresh-water systems. These relationships between the ecosystems lead to the concept of a *biome*.

A biome is a homogeneous ecological formation that exists over a large region as tundra or steppes. The biosphere comprises all of the Earth's biomes — the entirety of places where life is possible — from the highest mountains to the depths of the oceans.

Biomes correspond rather well to subdivisions distributed along the latitudes, from the equator towards the poles, with differences based on to the physical environment (for example, oceans or mountain ranges) and to the climate. Their variation is generally related to the distribution of species according to their ability to tolerate temperature and/or dryness. For example, one may find photosynthetic algae only in the *photic* part of the ocean (where light penetrates), while conifers are mostly found in mountains.

Though this is a simplification of more complicated scheme, latitude and altitude approximate a good representation of the distribution of biodiversity within the biosphere. Very generally, the richness of biodiversity (as well for animal than plant species) is decreasing most rapidly near the equator and less rapidly as one approaches the poles.

The biosphere may also be divided into ecozones, which are very well defined today and primarily follow the continental borders. The ecozones are themselves divided into ecoregions, though there is not agreement on their limits.

Ecosystem Productivity

In an ecosystem, the connections between species are generally related to food and their role in the food chain. There are three categories of organisms:

- *Producers* — usually plants which are capable of photosynthesis but could be other organisms such as bacteria around ocean vents that are capable of chemosynthesis.
- *Consumers* — animals, which can be primary consumers (herbivorous), or secondary or tertiary consumers (carnivorous and omnivores).
- *Decomposers* — bacteria, mushrooms which degrade organic matter of all categories, and restore minerals to the environment. And decomposers can also decompose decaying animals

These relations form sequences, in which each individual consumes the preceding one and is consumed by the one following, in what are called food chains or food network. In a food network, there will be fewer organisms at each level as one follows the links of the network up the chain.

These concepts lead to the idea of biomass (the total living matter in a given place), of primary productivity (the increase in the mass of plants during a given time) and of secondary productivity (the living matter produced by consumers and the decomposers in a given time).

These two last ideas are key, since they make it possible to evaluate the load capacity — the number of organisms which can be supported by a given ecosystem. In any food network, the energy contained in the level of the producers is not completely transferred to the consumers. And the higher one goes up the chain, the more energy and resources is lost and consumed. Thus, from an energy—and environmental—point of view, it is more efficient for humans to be primary consumers (to subsist from vegetables, grains, legumes, fruit, etc.) than as secondary consumers (from eating herbivores, omnivores, or their products, such as milk, chickens, cattle, sheep, etc.) and still more so than as a tertiary consumer (from consuming carnivores, omnivores, or their products, such as fur, pigs, snakes, alligators, etc.). An ecosystem(s) is unstable when the load capacity is overrun and is especially unstable when a population doesn't have an ecological niche and overconsumers.

The productivity of ecosystems is sometimes estimated by comparing three types of land-based ecosystems and the total of aquatic ecosystems:

- The forests (1/3 of the Earth's land area) contain dense biomasses and are very productive. The total production of the world's forests corresponds to half of the primary production.
- Savannas, meadows, and marshes (1/3 of the Earth's land area) contain less dense biomasses, but are productive. These ecosystems represent the major part of what humans depend on for food.
- Extreme ecosystems in the areas with more extreme climates — deserts and semi-deserts, tundra, alpine meadows, and steppes — (1/3 of the Earth's land area) have very sparse biomasses and low productivity
- Finally, the marine and fresh water ecosystems (3/4 of Earth's surface) contain very sparse biomasses (apart from the coastal zones).

Humanity's actions over the last few centuries have seriously reduced the amount of the Earth covered by forests (deforestation), and have increased agro-ecosystems (agriculture). In recent decades, an increase in the areas occupied by extreme ecosystems has occurred (desertification).

Ecological Crisis

Generally, an ecological crisis occurs with the loss of adaptive capacity when the resilience of an environment or of a species or a population evolves in a way unfavourable to coping with perturbations that interfere with that ecosystem, landscape or species survival.It may be that the environment quality degrades compared to the species needs, after a change in an abiotic ecological factor (for example, an increase of temperature, less significant rainfalls). It may be that the environment becomes unfavourable for the survival of a species (or a population) due to an increased pressure of predation (for example overfishing). Lastly, it may be that the situation becomes unfavourable to the quality of life of the species (or the population) due to a rise in the number of individuals (overpopulation).

Ecological crises vary in length and severity, occurring within a few months or taking as long as a few million years. They can also be of natural or anthropic origin. They may relate to one unique species or to many species, as in an Extinction event. Lastly, an ecological crisis may be local (as an oil spill) or global (a rise in the sea level due to global warming).

According to its degree of endemism, a local crisis will have more or less significant consequences, from the death of many individuals to the total extinction of a species. Whatever its origin, disappearance of one or several species often will involve a rupture in the food chain, further impacting the survival of other species.

In the case of a global crisis, the consequences can be much more significant; some extinction events showed the disappearance of more than 90% of existing species at that time. However, it should be noted that the disappearance of certain species, such as the dinosaurs, by freeing an ecological niche, allowed the development and the diversification of the mammals. An ecological crisis thus paradoxically favored biodiversity.

Sometimes, an ecological crisis can be a specific and reversible phenomenon at the ecosystem scale. But more generally, the crises impact will last. Indeed, it rather is a connected series of events, that occur till a final point. From this stage, no return to the previous stable state is possible, and a new stable state will be set up gradually (see homeorhesy).

Lastly, if an ecological crisis can cause extinction, it can also more simply reduce the quality of life of the remaining individuals. Thus, even if the diversity of the human population is sometimes considered threatened (see in particular indigenous people), few people envision human disappearance at short span. However, epidemic diseases, famines, impact on health of reduction of air quality, food crises, reduction of living space, accumulation of toxic or non degradable wastes, threats on keystone species (great apes, panda, whales) are also factors influencing the well-being of people.

Due to the increases in technology and a rapidly increasing population, humans have more influence on their own environment than any other ecosystem engineer.

Some common examples of ecological crises are:

- The Exxon Valdez oil spill off the coast of Alaska in 1989
- Permian-Triassic extinction event 250 million of years ago
- Cretaceous–Tertiary extinction event 65 million years ago
- Global warming related to the Greenhouse effect. Warming could involve flooding of the Asian deltas (see also eco refugees), multiplication of extreme

weather phenomena and changes in the nature and quantity of the food resources (see Global warming and agriculture). See also international Kyoto Protocol.

- Ozone layer hole issue
- Deforestation and desertification, with disappearance of many species.
- Volcanic eruptions such as Mount St. Helens and the Tunguska and other impact events
- The nuclear meltdown at Chernobyl in 1986 caused the death of many people and animals from cancer, and caused mutations in a large number of animals and people. The area around the plant is now abandoned by humans because of the large amount of radiation generated by the meltdown. Twenty years after the accident, the animals have returned.

ENVIRONMENTAL SCIENCE

Environmental science is the study of interactions among physical, chemical, and biological components of the environment. It is an interdisciplinary science overlapping the categories in Natural sciences, Engineering sciences and Social sciences. In nature, Environmental science focuses on pollution and degradation of the environment related to human activities and their impact on biodiversity and sustainability. As an interdisciplinary field, environmental science also applies knowledge from economics, law and social sciences. It can be applied to cyberspace environment also. In nature, Physics is used to understand the flux of material and energy interaction and construct mathematical models of environmental phenomena. Chemistry is applied to understand the molecular interactions among natural systems. Biology is fundamental to describing the effects within the plant and animal kingdoms.

Environmental science came alive as a substantive, active field of scientific investigation in the 1960s and 1970s. This was driven by the need for a large multi-disciplined team to analyze complex environmental problems, the arrival of substantive environmental laws requiring specific environmental protocols of investigation, and growing public awareness of a need for action in addressing environmental problems.

Environmental science encompasses issues such as climate change, conservation, biodiversity, groundwater and soil contamination, use of natural resources, waste management, sustainable development, air pollution and noise pollution. Due to the interdisciplinary nature of environmental science, teams of professionals commonly work together to conduct environmental research or to produce Environmental Impact Statements, as required by the U.S. National Environmental Policy Act (NEPA) or under state laws. Other professional organizations engender work in environmental science and aid in communication among the diverse sciences.

Sub-Categories

Atmospheric sciences examines the new phenomenology of the Earth's gaseous outer layer with emphasis upon interrelation to other systems. Atmospheric sciences comprises meteorological studies, greenhouse gas phenomena, atmospheric dispersion modeling of airborne contaminants,[1][2] sound propagation phenomena related to noise pollution, and even light pollution

Taking the example of the global warming phenomena, physicists create computer models of atmospheric circulation and infra-red radiation transmission, chemists examine the inventory of atmospheric chemicals and their reactions, biologists analyze the plant and animal contributions to carbon dioxide fluxes, and specialists such as meteorologists and oceanographers add additional breadth in understanding the atmospheric dynamics.

Ecology studies typically analyze the dynamics among an interrelated set of populations, or a population and some aspect of its environment. These studies could address endangered species, predator/prey interactions, habitat integrity, effects upon populations by environmental contaminants, or impact analysis of proposed land development upon species viability.

An interdisciplinary analysis of an ecological system which is being impacted by one or more stressors might include several related environmental science fields. For example one might examine an estuarine setting where a proposed industrial development could impact certain species by water pollution and air pollution. For this study biologists would describe the flora and fauna, chemists would analyze the transport of water pollutants to the marsh, physicists would calculate air pollution emissions and geologists would assist in understanding the marsh soils and bay muds.

Environmental chemistry is the study of chemical alterations in the environment. Principal areas of study include soil contamination and water pollution. The topics of analysis involve chemical degradation in the environment, multi-phase transport of chemicals (for example, evaporation of a solvent containing lake to yield solvent as an air pollutant), and chemical effects upon biota.

As an example study, consider the case of a leaking solvent tank which has entered the soil upgradient of a habitat of an endangered species of amphibian. Physicists would develop a computer model to understand the extent of soil contamination and subsurface transport of solvent, chemists would analyze the molecular bonding of the solvent to the specific soil type and biologists would study the impacts upon soil arthropods, plants and ultimately pond dwelling copepods who are the food of the endangered amphibian.

Geosciences include Environmental Geology, Hydrology, Physical Geography, Climatology and Geomorphology. It may also embrace oceanography and other related fields.

As an example study of soils erosion, calculations would be made of surface runoff by soil scientists. Hydrologists would assist in examining sediment transport in overland flow. Physicists would contribute by assessing the changes in light transmission in the receiving waters. Biologists would analyze subsequent impacts to aquatic flora and fauna from increases in water turbidity.

Environmental assessment is the process of appraisal through which environmental protection and sustainable development may be considered. Environmental assessments typically involve collection of field data, this can be from stakeholders and the ambient environment, and serves to harmonize the linkages between the different branches of the environment and development.

Environmental microbiology is the study of the composition and physiology of microbial communities in the environment. The environment in this case means the soil, water, air and sediments covering the planet and can also include the animals and plants that inhabit these areas. Environmental microbiology also includes the study of microorganisms that exist in artificial environments such as bioreactors.

Regulations Driving the Studies

In the U.S. the National Environmental Policy Act (NEPA) of 1969 set forth requirements for analysis of major projects in terms of specific environmental criteria. Numerous state laws have echoed these mandates, applying the principles to local scale actions. The upshot has been an explosion of documentation and study of environmental consequences before the fact of development actions.

One can examine the specifics of environmental science by reading examples of Environmental Impact Statements prepared under NEPA such as: *Wastewater treatment expansion options discharging into the San Diego/Tiajuana Estuary*, *Expansion of the San Francisco International Airport*, *Development of the Houston, Metro Transportation system*, *Expansion of the metropolitan Boston MBTA transit system*, and *Construction of Interstate 66 through Arlington, Virginia*.

In England and Wales the Environmental Agency (EA), (EA) formed in 1996 is the leading public body for protecting and improving the environment and enforces the regulations listed on the communities and local government site (formerly the office of the deputy prime minister) which help drive the study environmental science in the UK. The Agency was set up under the Environment Act 1995 as an independent body and works closely with UK Government to enforce the regulations.

Terminology

"Environmental science" and "ecology" are different fields of study, although there is some overlap due to the multidisciplinary nature of environmental science. *Ecology* is the study of the interrealtions of living organisms, whether at the population, community, or ecosystem level. In contrast, environmental science is a broad area of study including diverse areas such as meteorology, atmospheric chemistry, soil chemistry, water chemistry, pollution studies, and biological responses of systems to anthropogenic influence. water quality, air quality, and soil chemistry).

ENVIRONMENTAL CHEMISTRY

Environmental chemistry is the scientific study of the chemical and biochemical phenomena that occur in natural places. It should not be confused with green chemistry, which seeks to reduce potential pollution at source. It can be defined as the study of the sources, reactions, transport, effects, and fates of chemical species in the air, soil, and water environments; and the effect of human activity on these. Environmental chemistry is an interdisciplinary science that includes atmospheric, aquatic and soil chemistry, as well as heavily relying on analytical chemistry and being related to environmental and other areas of science.

Environmental chemistry involves first understanding how the uncontaminated environment works, which chemicals in what concentrations are present naturally, and with what effects. Without this it would be impossible to accurately study the effects humans have on the environment through the release of chemicals.

Environmental chemists draw on a range of concepts from chemistry and various environmental sciences to assist in their study of what is happening to a chemical species in the environment. Important general concepts from chemistry include understanding chemical reactions and equations, solutions, units, sampling, and analytical techniques

Contamination

A contaminant is a substance present in nature due to human activity, that would not otherwise be there. The term contaminant is often used interchangeably with *pollutant*, which is a substance that has a detrimental impact on the environment it is in. Whilst a contaminant is sometimes defined as a substance present in the environment as a result of human activity, but without harmful effects, it is sometimes the case that toxic or harmful effects from contamination only become apparent at a later date.

The "medium" (e.g. soil) or organism (e.g. fish) affected by the pollutant or contaminant is called a *receptor*, whilst a *sink* is a chemical medium or species that retains and interacts with the pollutant.

Environmental Indicators

Chemical measures of water quality include dissolved oxygen (DO), chemical oxygen demand (COD), biological oxygen demand (BOD), and pH.

Applications

Environmental chemistry is used by the Environment Agency (in England and Wales), the Environmental Protection Agency (in the United States), and other environmental agencies and research bodies around the world to detect and identify the nature and source of pollutants. These can include:

- Heavy metal contamination of land by industry. These can then be transported into water flows and be taken up by living organisms.
- Nutrients such as nitrate and phosphate leaching from agricultural land into water courses, which can lead to algal blooms and eutrophication.

ENVIRONMENTAL IMPACT ASSESSMENT

An Environmental Impact Statement (EIS) is an assessment of the likely influence a project may have on the environment. "Environmental Impact Assessment can be defined as: The process of identifying, predicting, evaluating and mitigating the biophysical, social, and other relevant effects of development proposals prior to major decisions being taken and commitments made." (IAIA 1999). The purpose of the assessment is to ensure that decision-makers consider environmental impacts before deciding whether to proceed with new projects.

Overview

The US Environmental Protection Agency pioneered the use of pathway analysis to determine the likely human health impact of environmental factors. The technology for performing such analysis is properly labelled *environmental science*. The principal phenomena or pathways of impact are: soil contamination impacts, air pollution impacts, noise health effects, ecology impacts including endangered species assessment, geological hazards assessment and water pollution impacts. Pathway analysis and The Natural Step definitions subsequently became the basis of the global ISO 14000 series of environmental management standards and the more recent ISO 19011 auditing standard; however, these ISO standards are not in common use in the U.S. and most other countries.

After an EIA analysis, the Precautionary Principle and Polluter Pays may be applied to prevent, limit, or require strict liability or insurance coverages to a project, based on its likely harms.

Environmental impact analysis is sometimes controversial and contested. Related analysis of social impacts is achieved by Social impact assessment. Analysis of business impacts is achieved by Context analysis. Design impacts are assessed in relation to Context theory.

EIS predicts what a specific action can do to the environment.

EIA Around the World

China

The Environmental Impact Assessment Law (EIA Law) requires an environmental impact assessment to be completed prior to project construction. However, if a developer completely ignores this requirement and builds a project without submitting an environmental impact statement, the only penalty is that the environmental protection bureau (EPB) may require the developer to do a make-up environmental assessment. If the developer does not complete this make-up assessment within the designated time, only then is the EPB authorized to fine the developer. Even so, the possible fine is capped at a maximum of about US$25,000, a fraction of the overall cost of most major projects. The lack of more stringent enforcement mechanisms has resulted in a significant percentage of projects not completing legally required environmental impact assessments prior to construction.

China's State Environmental Protection Administration (SEPA) used the legislation to halt 30 projects in 2004, including three hydro-power plants under the Three Gorges Project Company. Although one month later (Note as a point of reference, that the typical EIA for a major project in the USA takes one to two years.), most of the 30 halted projects resumed their construction, reportedly having passed the environmental assessment, the fact that these key projects' construction was ever suspended was notable.

A joint investigation by SEPA and the Ministry of Land and Resources in 2004 showed that 30 to 40 per cent of the mining construction projects went through the procedure of environment impact assessment as required, while in some areas only 6 to 7 per cent did so. This partly explains why China has witnessed so many mining accidents in recent years.

SEPA alone cannot guarantee the full enforcement of environmental laws and regulations, observed Professor Wang Canfa, director of the centre to help environmental victims at China University of Political Sciences and Law. In fact, according to Wang, the rate of China's environmental laws and regulations that are actually enforced is estimated to be barely 10 per cent.

EU

The EIA Directive on Environmental Impact Assessment of the effects of projects on the environment was first introduced in 1985 and was amended in 1997. The directive was amended again in 2003 following the 1998 signature by the EU of the Aarhus Convention on public participation in environmental matters. The issue was enlarged to the assessment of plans and programmes by the so called *SEA-Directive* in 2001 which is now in force and establishes a mix of mandatory and discretionary procedures for assessing environmental impacts.

Under the EU directive, an EIA must provide certain information to comply. There are seven key areas that are required:

1. Description of the project
    - Description of actual project and site description
    - Break the project down into its key components, ie construction, operations, decommissioning
    - For each component list all of the sources of environmental disturbance
    - For each component all the inputs and outputs must be listed, eg, air pollution, noise, hydrology
2. Alternatives that have been considered
    - Examine alternatives that have been considered
    - Eg - In a biomass power station, will the fuel be sourced locally or nationally?
3. Description of the environment
    - List of all aspects of the environment that may be effected by the development
    - eg populations, fauna, flora, air, soil, water, humans, landscape, cultural heritage
    - This section is best carried out with the help of local experts, eg the RSPB in the UK
4. Description of the significant effects on the environment
    - The word significant is crucial here as the definition can vary
    - 'Significant' needs to be defined

- The most frequent method used here is use of the Leopold matrix
- The maxtrix is a tool used in the systematic examination of potential interactions
- Eg In a windfarm development a significant impact may be collisions with birds

5. Mitigation
   - This is where EIA is most useful
   - Once section 4 has been completed it will be obvious where the impacts will be greatest
   - Using this information ways to avoid negative impacts should be developed
   - Best working with the developer with this section as they know the project best
   - Using the windfarm example again construction could be out of bird nesting seasons
6. Non-technical summary
   - The EIA will be in the public domain and be used in the decision making process
   - It is important that the information is available to the public
   - This section is a summary that does not include jargon or complicated diagrams
   - It should be understood by the informed lay-person
7. Lack of know-how/technical difficulties
   - This section is to advise any areas of weakness in knowledge
   - It can be used to focus areas of future research
   - Some developers see the EIA as a starting block for good environmental management

South Africa

The Nelson Mandela Metropolitan University offers a full course of third year on Environmental Impact Assessment in South Africa.

New Zealand

In New Zealand, EIA is usually referred to as *Assessment of Environmental Effects* (AEE). The first use of EIA's dates back to a Cabinet minute passed in 1974 called Environmental Protection and Enhancement Procedures. This had no legal force and only related to the activities of government departments. When the Resource Management Act was passed in 1991, an EIA was required as part of a resource consent application. Section 88 of the Act spells this out.

United States

Under United States environmental law an EIA is referred to as the *Environmental Impact Statement* (EIS), and originated in the National Environmental Policy Act (NEPA), enacted in the United States in 1969. Certain actions of federal agencies must be preceded by an EIS. Contrary to a widespread misconception, NEPA does not prohibit the federal government or its licensees/permittees from harming the environment, nor does it specify any penalty if the EIS turns out to be inaccurate, intentionally or otherwise. NEPA requires that plausible statements as to the prospective impacts be disclosed in advance. The purpose of NEPA process is to ensure that the decision maker is fully informed of the environmental aspects and consequences prior to making the final decision.

Usually, an agency will release a *Draft Environmental Impact Statement* (DEIS) for comment. Interested parties and the general public have the opportunity to comment on the draft, after which the agency will approve the *Final Environmental Impact Statement* (FEIS). Occasionally, the agency will later release a *Supplemental Environmental Impact Statement* (SEIS), especially if environmental conditions suddenly change after the issuance of the FEIS.

The adequacy of an EIS can be challenged in court. Major proposed projects have been blocked because of an agency's failure to prepare an acceptable EIS. One prominent example was the Westway landfill and highway development in and along the Hudson River in New York City . Another prominent case involved the Sierra Club suing the Nevada Department of Transportation over its denial of Sierra Club's request to issue a supplemental EIS addressing air emissions of particulate matter and hazardous air pollutants in the case of widening US Highway 95 through Las Vegas. The case reached the 9th Circuit Court of the United States, which led to construction on the highway being halted until the court's final decision. The case was settled prior to the court's final decision.

Several US state governments that have adopted "little NEPA's," i.e., state laws imposing EIS requirements for particular state actions and some of those state laws refer to the required environmental impact studies as Environmental Impact Reports or Environmental Impact Assessments. [4] For example, the California Environmental Quality Act (CEQA) requires an *Environmental Impact Report* (EIR).

These various state requirements are yielding voluminous data not just upon impacts of individual projects, but also to elucidate scientific areas that had not been sufficiently researched. For example, in a seemingly routine *Environmental Impact Report* for the city of Monterey, California, information came to light that led to the official federal endangered species listing of Hickman's potentilla, a rare coastal wildflower.

ENVIRONMENTAL PRESERVATION

Environmental preservation is the strict setting aside of natural resources to prevent the use or contact by humans or by human intervention. In terms of policy making this often means setting aside areas as nature reserves (otherwise known as wildlife reserves), parks, or other conservation areas. These areas usually restrict or prohibit activities that may cause damage to habitat or wildlife. Such activities may include: logging, hunting, fishing, mining and so on. Just as often legal devices such as laws and regulations may be employed, such as the Endangered Species Act in the United States, which is not dependent on designating a specific geographic area aside for conservation. A government's environmental policy will determine which areas or species, etc. are protected and how.

Environmental preservation is different from *conservation*; *conservation* allows for sustainable development, whereas *preservation* is complete restriction.

3

Atmospheric Science and Earth Science

ATMOSPHERIC SCIENCES

Atmospheric sciences is an umbrella term for the study of the atmosphere, its processes, the effects other systems have on the atmosphere, and the effects of the atmosphere on these other systems. Meteorology includes atmospheric chemistry and atmospheric physics with a major focus on weather forecasting. Climatology is the study of atmospheric changes (both long and short-term) that define average climates and their change over time, due to both natural climate variability and anthropogenic climate variability. Atmospheric science has been extended to the field of planetary science and the study of the atmospheres of the planets of the solar system.

Atmospheric Chemistry

Atmospheric chemistry is a branch of atmospheric science in which the chemistry of the Earth's atmosphere and that of other planets is studied. It is a multidisciplinary field of research and draws on environmental chemistry, physics, meteorology, computer modeling, oceanography, geology and volcanology and other disciplines. Research is increasingly connected with other areas of study such as climatology.

The composition and chemistry of the atmosphere is of importance for several reasons, but primarily because of the interactions between the atmosphere and living organisms. The composition of the Earth's atmosphere has been changed by human activity and some of these changes are harmful to human health, crops and ecosystems. Examples of problems which have been addressed by atmospheric chemistry include acid rain, photochemical smog and global warming. Atmospheric chemistry seeks to understand the causes of these problems, and by obtaining a theoretical understanding of them, allow possible solutions to be tested and the effects of changes in government policy evaluated.

Atmospheric Dynamics

Atmospheric dynamics involves the study of observations and theory dealing with all motion systems of meteorological importance. The list includes diverse phenomena as thunderstorms, tornadoes, gravity waves, tropical cyclones, extratropical cyclones, jet streams, and global-scale circulations. The goal of dynamical studies is to explain the observed circulations on the basis of fundamental principles from physics. The objectives of such studies include improving weather forecasting, developing methods for predicting seasonal and interannual climate fluctuations, and understanding the implications of human-induced perturbations (e.g., increased carbon dioxide concentrations or depletion of the ozone layer) on the global climate.

Atmospheric Physics

Atmospheric physics is the application of physics to the study of the atmosphere. Atmospheric physicists attempt to model Earth's atmosphere and the atmospheres of the other planets using fluid flow equations, chemical models, radiation balancing, and energy transfer processes in the atmosphere and underlying oceans. In order to model weather systems, atmospheric physicists employ elements of scattering theory, wave propagation models, cloud physics, statistical mechanics and spatial statistics which are highly mathematical and related to physics. It has close links to meteorology and climatology and also covers the design and construction of instruments for studying the atmosphere and the interpretation of the data they provide, including remote sensing instruments.

In the UK, atmospheric studies are underpinned by the Meteorological Office. Divisions of the U.S National Oceanic and Atmospheric Administration (NOAA) oversee research projects and weather modeling involving atmospheric physics. The US National Astronomy and Ionosphere Center also carries out studies of the high atmosphere.

Climatology

In contrast to meteorology, which studies short term weather systems lasting up to a few weeks, climatology studies the frequency and trends of those systems. It studies the periodicity of weather events over years to millennia, as well as changes in long-term average weather patterns, in relation to atmospheric conditions. Climatologists, those who practice climatology, study both the nature of climates - local, regional or global - and the natural or human-induced factors that cause climates to change. Climatology considers the past and can help predict future climate change.

Phenomena of climatological interest include the atmospheric boundary layer, circulation patterns, heat transfer (radiative, convective and latent), interactions between the atmosphere and the oceans and land surface (particularly vegetation, land use and topography), and the chemical and physical composition of the atmosphere. Related disciplines include astrophysics, atmospheric physics, chemistry, ecology, geology, geophysics, glaciology, hydrology, oceanography, and volcanology.

AIR POLLUTION

Air pollution is the modification of the natural characteristics of the atmosphere by a chemical, particulate matter, or biological agent. The atmosphere is a complex, dynamic natural gaseous system that is essential to support life on planet Earth. Stratospheric ozone depletion due to air pollution has long been recognized as a threat to human health as well as to the Earth's ecosystems.

Worldwide air pollution is responsible for large numbers of deaths and cases of respiratory disease. While major stationary sources are often identified with air pollution, the greatest source of emissions is actually mobile sources, mainly automobiles. Gases such as carbon dioxide, which contribute to global warming, have recently gained recognition as pollutants by climate scientists, while they also recognize that carbon dioxide is essential for plant life through photosynthesis.

Pollutants

There are many substances in the air which may impair the health of plants and animals (including humans), or reduce visibility. These arise both from natural processes and human activity. Substances not naturally found in the air or at greater concentrations or in different locations from usual are referred to as *pollutants*.

Pollutants can be classified as either primary or secondary. Primary pollutants are substances directly emitted from a process, such as ash from a volcanic eruption or the carbon monoxide gas from a motor vehicle exhaust.

Secondary pollutants are not emitted directly. Rather, they form in the air when primary pollutants react or interact. An important example of a secondary pollutant is ground level ozone - one of the many secondary pollutants that make up photochemical smog.

Note that some pollutants may be both primary and secondary: that is, they are both emitted directly and formed from other primary pollutants.

Major primary pollutants produced by human activity include:

- Sulfur oxides (SO_x) especially sulfur dioxide are emitted from burning of coal and oil.
- Nitrogen oxides (NO_x) especially nitrogen dioxide are emitted from high temperature combustion. Can be seen as the brown haze dome above or plume downwind of cities.
- Carbon monoxide is colourless, odourless, non-irritating but very poisonous gas. It is a product by incomplete combustion of fuel such as natural gas, coal or wood. Vehicular exhaust is a major source of carbon monoxide.
- Carbon dioxide (CO_2), a greenhouse gas emitted from combustion.
- Volatile organic compounds (VOC), such as hydrocarbon fuel vapors and solvents.
- Particulate matter (PM), measured as smoke and dust. PM_{10} is the fraction of suspended particles 10 micrometers in diameter and smaller that will enter the nasal cavity. $PM_{2.5}$ has a maximum particle size of 2.5 μm and will enter the bronchies and lungs.
- Toxic metals, such as lead, cadmium and copper.
- Chlorofluorocarbons (CFCs), harmful to the ozone layer emitted from products currently banned from use.
- Ammonia (NH_3) emitted from agricultural processes.
- Odors, such as from garbage, sewage, and industrial processes
- Radioactive pollutants produced by nuclear explosions and war explosives, and natural processes such as radon.

Secondary pollutants include:

- Particulate matter formed from gaseous primary pollutants and compounds in photochemical smog, such as nitrogen dioxide.
- Ground level ozone (O_3) formed from NOx and VOCs.
- Peroxyacetyl nitrate (PAN) similarly formed from NOx and VOCs.

Minor air pollutants include:

- A large number of minor hazardous air pollutants. Some of these are regulated in USA under the Clean Air Act and in Europe under the Air Framework Directive.
- A variety of persistent organic pollutants, which can attach to particulate matter.

Sources of Air Pollution

Sources of air pollution refer to the various locations, activities or factors which are responsible for the releasing of pollutants in the atmosphere. These sources can be classified into two major categories which are:

Anthropogenic sources (human activity) mostly related to burning different kinds of fuel

- "Stationary Sources" as smoke stacks of power plants, manufacturing facilities, municipal waste incinerators
- "Mobile Sources" as motor vehicles, aircraft etc.
- Combustion-fired power plants
- Controlled burn practices used in agriculture and forestry management
- Motor vehicles generating air pollution emissions.
- Marine vessels, such as container ships or cruise ships, and related port air pollution.
- Burning wood, fireplaces, stoves, furnaces and incinerators
- Oil refining, power plant operation and industrial activity in general.
- Chemicals, dust and crop waste burning in farming, (see Dust Bowl).
- Fumes from paint, hair spray, varnish, aerosol sprays and other solvents.
- Waste deposition in landfills, which generate methane.
- Military uses, such as nuclear weapons, toxic gases, germ warfare and rocketry.

Natural Sources

- Dust from natural sources, usually large areas of land with little or no vegetation.
- Methane, emitted by the digestion of food by animals, for example cattle.
- Radon gas from radioactive decay within the Earth's crust.
- Smoke and carbon monoxide from wildfires.
- Volcanic activity, which produce sulfur, chlorine, and ash particulates.

Indoor Air Quality (IAQ)

A lack of ventilation indoors concentrates air pollution where people often spend the majority of their time. Radon (Rn) gas, a carcinogen, is exuded from the Earth in certain locations and trapped inside houses. Researchers have found that radon gas is responsible for over 1,800 deaths annually in the United Kingdom. Building materials including carpeting and plywood emit formaldehyde (H_2CO) gas. Paint and solvents give off volatile organic compounds (VOCs) as they dry. Lead paint can degenerate into dust and be inhaled. Intentional air pollution is introduced with the use of air fresheners, incense, and other scented items. Controlled wood fires in stoves and fireplaces can add significant amounts of smoke particulates into the air, inside and out. Indoor pollution fatalities may be caused by using pesticides and other chemical sprays indoors without proper ventilation.

Carbon monoxide (CO) poisoning and fatalities are often caused by faulty vents and chimneys, or by the burning of charcoal indoors. 56,000 Americans died from CO in the period 1979-1988. Chronic carbon monoxide poisoning can result even from poorly adjusted pilot lights. Traps are built into all domestic plumbing to keep sewer gas, hydrogen sulfide, out of interiors. Clothing emits tetrachloroethylene, or other dry cleaning fluids, for days after dry cleaning.

Though its use has now been banned in many countries, the extensive use of asbestos in industrial and domestic environments in the past has left a potentially very dangerous material in many localities. Asbestosis is a chronic inflammatory medical condition affecting the tissue of the lungs. It occurs after long-term, heavy exposure to asbestos from asbestos-containing materials in structures. Sufferers have severe dyspnea (shortness of breath) and are at an increased risk regarding several different types of lung cancer. As clear explanations are not always stressed in non-technical literature, care should be taken to distinguish between several forms of relevant diseases. According to the World Health Organisation (WHO), these may defined as; asbestosis, *lung cancer*, and *mesothelioma* (generally a very rare form of cancer, when more widespread it is almost always associated with prolonged exposure to asbestos).

Biological sources of air pollution are also found indoors, as gases and airborne particulates. Pets produce dander, people produce dust from minute skin flakes and decomposed hair, dust mites in bedding, carpeting and furniture produce enzymes and micron-sized fecal droppings, inhabitants emit methane, mold forms in walls and generates mycotoxins and spores, air conditioning systems can incubate Legionnaires' disease and mold, and houseplants, soil and surrounding gardens can produce pollen, dust, and mold. Indoors, the lack of air circulation allows these airborne pollutants to accumulate more than they would otherwise occur in nature.

Health Effects

The World Health Organization states that 2.4 million people die each year from causes directly attributable to air pollution. Many of these mortalities are attributable to indoor air pollution. Worldwide more deaths per year are linked to air pollution than to automobile accidents. Published in 2005 suggests that 310,000 Europeans die from air pollution annually. Direct causes of air pollution related deaths include aggravated asthma, bronchitis, emphysema, lung and heart diseases, and respiratory allergies. The US EPA estimates that a proposed set of changes in diesel engine technology (*Tier 2*) could result in 12,000 fewer *premature mortalities*, 15,000 fewer heart attacks, 6,000 fewer emergency room visits by children with asthma, and 8,900 fewer respiratory-related hospital admissions each year in the United States.

The worst short term civilian pollution crisis in India was the 1984 Bhopal Disaster. Leaked industrial vapors from the Union Carbide factory, belonging to Union Carbide, Inc., U.S.A., killed more than 2,000 people outright and injured anywhere from 150,000 to 600,000 others, some 6,000 of whom would later die from their injuries. The United Kingdom suffered its worst air pollution event when the December 4th Great Smog of 1952 formed over London. In six days more than 4,000 died, and 8,000 more died within the following months. An accidental leak of anthrax spores from a biological warfare laboratory in the former USSR in 1979 near Sverdlovsk is believed to have been the cause of hundreds of civilian deaths. The worst single incident of air pollution to occur in the United States of America occurred in Donora, Pennsylvania in late October, 1948, when 20 people died and over 7,000 were injured.

The health effects caused by air pollutants may range from subtle biochemical and physiological changes to difficulty in breathing, wheezing, coughing and aggravation of existing respiratory and cardiac conditions. These effects can result in increased medication use, increased doctor or emergency room visits, more hospital admissions and premature death. The human health effects of poor air quality are far reaching, but principally affect the body's respiratory system and the cardiovascular system. Individual reactions to air pollutants depend on the type of pollutant a person is exposed to, the degree of exposure, the individual's health status and genetics. People who exercise outdoors, for example, on hot, smoggy days increase their exposure to pollutants in the air.

Reduction Efforts

There are many air pollution control technologies and urban planning strategies available to reduce air pollution; however, worldwide costs of addressing the issue are high. Of course, these costs are a small fraction of the economic damage that air pollution will inflict on every nation of earth. Within the last decade the cost of air pollution annually in most of Europe is between 1-3 percent GDP and is at least 5 percent GDP of China.

Many countries have programs to or are debating how to reduce dependence on fossil fuels for energy production and shift toward renewable energy technologies or nuclear power plants.

Efforts to reduce pollution from mobile sources includes primary regulation (many developing countries have permissive regulations), expanding regulation to new sources (such as cruise and transport ships, farm equipment, and small gas-powered equipment such as lawn trimmers, chainsaws, and snowmobiles), increased fuel efficiency (such as through the use of hybrid vehicles), conversion to cleaner fuels (such as bioethanol, [[biodiesel), or conversion to electric vehicles with renewable energy sources (batteries or clean fuel such as hydrogen being used for transport and storage).

Control Devices

The following items are commonly used as pollution control devices by industry or transportation devices. They can either destroy contaminants or remove them from an exhaust stream before it is emitted into the atmosphere.

- Particulate control
  - Mechanical collectors (dust cyclones, multicyclones)
  - Electrostatic precipitators
  - Baghouses
  - Particulate scrubbers
- Scrubbers
  - Baffle spray scrubber
  - Cyclonic spray scrubber
  - Ejector venturi scrubber
  - Mechanically aided scrubber
  - Spray tower
  - Wet scrubber
- NOx control
  - Low NOx burners
  - Selective catalytic reduction (SCR)
  - Selective non-catalytic reduction (SNCR)
  - NOx scrubbers
  - Exhaust gas recirculation
  - Catalytic converter (also for VOC control)

- VOC abatement
  - Adsorption systems, such as activated carbon
  - Flares
  - Thermal oxidizers
  - Catalytic oxidizers
  - Biofilters
  - Absorption (scrubbing)
  - Cryogenic condensers
  - Vapor recovery systems
- Acid Gas/SO_2 control
  - Wet scrubbers
  - Dry scrubbers
  - Flue gas desulfurization
- Mercury control
  - Sorbent Injection Technology
  - Electro-Catalytic Oxidation (ECO)
  - K-Fuel
- Dioxin and furan control
- Miscellaneous associated equipment
  - Source capturing systems
  - Continuous emissions monitoring systems (CEMS)

Air Quality Standards

In general, there are two types of air quality standards. The first class of standards (such as the U.S. National Ambient Air Quality Standards) set maximum atmospheric concentrations for specific pollutants. Environmental agencies enact regulations which are intended to result in attainment of these target levels. The second class (such as the North American Air Quality Index) take the form of a scale with various thresholds, which is used to communicate to the public the relative risk of outdoor activity. The scale may or may not distinguish between different pollutants.

Canada

In Canada, air quality is typically evaluated against standards set by the Canadian Council of Minister for the Environment (CCME), an inter-governmental body of federal, provincial and territorial Ministers responsible for the environment. The CCME has set Canada Wide Standards(CWS). These are:

- CWS for PM2.5 = 30 μg/m3 (24 hour averaging time, by year 2010, based on 98th percentile ambient measurement annually, averaged over 3 consecutive years).
- CWS for ozone = 65 ppb (8-hour averaging time, by year 2010, achievement is based on the 4th highest measurement annually, averaged over 3 consecutive years.

Note that there is no consequence in Canada to not achieving these standards. In addition, these only apply to jurisdictions with populations greater than 100,000. Further, provinces and territories may set more stringent standards than those set by the CCME.

European Union

National Emission Ceilings (NEC) for certain atmospheric pollutants are regulated by Directive 2001/81/EC (NECD). As part of the preparatory work associated with the revision of the NECD, the European Commission is assisted by the NECPI working group (National Emission Ceilings – Policy Instruments).

United Kingdom

Air quality targets set by the UK's Department for Environment, Food and Rural Affairs (DEFRA) are mostly aimed at local government representatives responsible for the management of air quality in cities, where air quality management is the most urgent. The UK has established an air quality network where levels of the key air pollutants are published by monitoring centers. Air quality in Oxford, Bath and London is particularly poor. One controversial study performed by the Calor Gas company and published in the Guardian newspaper compared walking in Oxford on an average day to smoking over sixty light cigarettes.

More precise comparisons can be collected from the UK Air Quality Archive which allows the user to compare a cities management of pollutants against the national air quality objectives set by DEFRA in 2000.

Localized peak values are often cited, but average values are also important to human health. The UK National Air Quality Information Archive offers almost real-time monitoring of "current maximum" air pollution measurements for many UK towns and cities. This source offers a wide range of constantly updated data, including:

- Hourly Mean Ozone ($\mu g/m^3$)

- Hourly Mean Nitrogen dioxide ($\mu g/m^3$)
- Maximum 15-Minute Mean Sulphur dioxide ($\mu g/m^3$)
- 8-Hour Mean Carbon monoxide (mg/m^3)
- 24-Hour Mean PM_{10} ($\mu g/m^3$ Grav Equiv)

DEFRA acknowledges that air pollution has a significant effect on health and has produced a simple banding index system is used to create a daily warning system that is issued by the BBC Weather Service to indicate air pollution levels. DEFRA has published guidelines for people suffering from respiratory and heart diseases.

United States

In the 1960s, 70s, and 90s, the United States Congress enacted a series of Clean Air Acts which significantly strengthened regulation of air pollution. Individual U.S. states, some European nations and eventually the European Union followed these initiatives. The Clean Air Act sets numerical limits on the concentrations of a basic group of air pollutants and provide reporting and enforcement mechanisms.

In 1999, the United States EPA replaced the Pollution Standards Index (PSI) with the Air Quality Index (AQI) to incorporate new PM2.5 and Ozone standards.

The effects of these laws have been very positive. In the United States between 1970 and 2006, citizens enjoyed the following reductions in annual pollution emissions:

- carbon monoxide emissions fell from 197 million tons to 89 million tons
- nitrogen oxide emissions fell from 27 million tons to 19 million tons
- sulfur dioxide emissions fell from 31 million tons to 15 million tons
- particulate emissions fell by 80%
- lead emissions fell by more than 98%

In an October 2006 letter to EPA, the agency's independent scientific advisors warned that the ozone smog standard "needs to be substantially reduced" and that there is "no scientific justification" for retaining the current, weaker standard. The scientists unanimously recommended a smog threshold of 60 to 70 ppb after they conducted an extensive review of the evidence.

The EPA has proposed, in June 2007, a new threshold of 75 ppb. This falls short of the scientific recommendation, but is an improvement over the current standard.

Polluting industries are lobbying to keep the current (weaker) standards in place. Environmentalists and public health advocates are mobilizing to support compliance with the scientific recommendations.

The National Ambient Air Quality Standards are pollution thresholds which trigger mandatory remediation plans by state and local governments, subject to enforcement by the EPA.

An outpouring of dust layered with man-made sulfates, smog, industrial fumes, carbon grit, and nitrates is crossing the Pacific Ocean on prevailing winds from booming Asian economies in plumes so vast they alter the climate. Almost a third of the air over Los Angeles and San Francisco can be traced directly to Asia. With it comes up to three-quarters of the black carbon particulate pollution that reaches the West Coast.

Geographic Areas Affected by Air Pollution

Cities

Air pollution is usually concentrated in densely populated metropolitan areas, especially in developing countries where environmental regulations are generally relatively lax. However, even populated areas in developed countries attain unhealthy levels of pollution.

Most Polluted World Cities by PM

Particulatematter,µg / m³ (2004)	*City*
169	Cairo, Egypt
150	Delhi, India
128	Kolkata, India (Calcutta)
125	Taiyuan, China
123	Chongqing, China
109	Kanpur, India
109	Lucknow, India
104	Jakarta, Indonesia
101	Shenyang, China

Countries

The source of these data is the Carbon Monitoring for Action (CARMA) database produced by the Center for Global Development.

Total Carbon Dioxide Emissions

10^6 Tons of CO_2 per year:

- United States: 2,790
- China: 2,680
- Russia: 661
- India: 583
- Japan: 400
- Germany: 356
- Australia: 226
- South Africa: 222
- United Kingdom: 212
- South Korea: 185

Per Capita Carbon Dioxide Emissions

Tons of CO_2 per year per capita:

- Australia: 10
- United States: 8.2
- United Kingdom: 3.2
- China: 1.8
- India: 0.5

Atmospheric Dispersion Modeling

The basic technology for analyzing air pollution is through the use of a variety of mathematical models for predicting the transport of air pollutants in the lower atmosphere. The principal methodologies are:

- Point source dispersion, used for industrial sources.
- Line source dispersion, used for airport and roadway air dispersion modeling
- Area source dispersion, used for forest fires or duststorms
- Photochemical models, used to analyze reactive pollutants that form smog

The point source problem is the best understood, since it involves simpler mathematics and has been studied for a long period of time, dating back to about the year 1900. It uses a Gaussian dispersion model for buoyant pollution plumes to forecast the air pollution isopleths, with consideration given to wind velocity, stack height, emission rate and stability class (a measure of atmospheric turbulence). This model has been extensively validated and calibrated with experimental data for all sorts of atmospheric conditions.

The roadway air dispersion model was developed starting in the late 1950s and early 1960s in response to requirements of the National Environmental Policy Act and the U.S. Department of Transportation (then known as the Federal Highway Administration) to understand impacts of proposed new highways upon air quality, especially in urban areas. Several research groups were active in this model development, among which were: the Environmental Research and Technology (ERT) group in Lexington, Massachusetts, the ESL Inc. group in Sunnyvale, California and the California Air Resources Board group in Sacramento, California. The research of the ESL group received a boost with a contract award from the United States Environmental Protection Agency to validate a line source model using sulfur hexafluoride as a tracer gas. This program was successful in validating the line source model developed by ESL inc. Some of the earliest uses of the model were in court cases involving highway air pollution, the Arlington, Virginia portion of Interstate 66 and the New Jersey Turnpike widening project through East Brunswick, New Jersey.

Area source models were developed in 1971 through 1974 by the ERT and ESL groups, but addressed a smaller fraction of total air pollution emissions, so that their use and need was not as widespread as the line source model, which enjoyed hundreds of different applications as early as the 1970s. Similarly photochemical models were developed primarily in the 1960s and 1970s, but their use was more specialized and for regional needs, such as understanding smog formation in Los Angeles, California.

Greenhouse Effect and Ocean Acidification

The greenhouse effect is a phenomenon whereby greenhouse gases, create a condition in the upper atmosphere causing a trapping of heat and leading to increased surface and lower tropospheric temperatures. The effect prevents the planet from severe cooling, and so benefits all living things. It shares this property with many other gases, the largest overall forcing on Earth coming from water vapour. Other greenhouse gases include methane, hydrofluorocarbons, perfluorocarbons, chlorofluorocarbons, NOx, and ozone. Many greenhouse gases, contain carbon, and some of that from fossil fuels.

This effect has been understood by scientists for about a century, and technological advancements during this period have helped increase the breadth and depth of data relating to the phenomenon. Currently, scientists are studying the role of changes in composition of greenhouse gases from natural and anthropogenic sources for the effect on climate change.

A number of studies have also investigated the potential for long-term rising levels of atmospheric carbon dioxide to cause slight increases in the acidity of ocean waters and the possible effects of this on marine ecosystems. However, carbonic acid is a very weak acid, and is utilized by marine organisms during photosynthesis.

EARTH SCIENCE

Earth science (also known as geoscience, the geosciences or the Earth Sciences), is an all-embracing term for the sciences related to the planet Earth. It is arguably a special case in planetary science, the Earth being the only known life-bearing planet. There are both reductionist and holistic approaches to Earth science. The major historic disciplines use physics, geology, geography, meteorology, mathematics, chemistry and biology to build a quantitative understanding of the principal areas or *spheres* of the Earth system.

Earth's Spheres

Earth science generally recognizes 4 spheres, the lithosphere, the hydrosphere, the atmosphere, and the biosphere. These correspond to rocks, water, air, and life. Some practitioners include the cryosphere (ice) as a distinct portion of the hydrosphere and the pedosphere (soil) as an active, intermixed sphere as part of Earth's spheres.

- Geology describes the rocky parts of the Earth's crust (or lithosphere) and its historic development. Major subdisciplines are mineralogy and petrology, geochemistry, geomorphology, paleontology, stratigraphy, structural geoogy, engineering geology and sedimentology.[3][4]
- Geophysics and Geodesy investigate the figure of the Earth, its reaction to forces and its magnetic and gravity fields. Geophysicists explore the Earth's core and mantle as well as the tectonic and seismic activity of the lithosphere.
- Soil science covers the outermost layer of the Earth's crust that is subject to soil formation processes (or pedosphere. Major subdisciplines include edaphology and pedology.
- Oceanography and hydrology (incl. limnology) describe the marine and freshwater domains of the watery parts of the Earth (or hydrosphere). Major subdisciplines include hydrogeology and physical, chemical, and biological oceanography.
- Glaciology covers the icy parts of the Earth (or cryosphere).
- Atmospheric sciences cover the gaseous parts of the Earth (or atmosphere) between the surface and the exosphere (~1000 km). Major subdisciplines are meteorology, climatology, atmospheric chemistry and atmospheric physics.
- A very important linking sphere is the biosphere, the study of which is biology. The biosphere consists of all forms of life, from single-celled organisms to pine trees to people. The interactions of Earth's other spheres - lithosphere/

geosphere, hydrosphere, atmosphere and/or cryosphere and pedosphere - create the conditions that can support life.

Earth's Interior

In geology, plate tectonics, mountain ranges, volcanoes, and earthquakes are phenomena that can be explained in terms of energy transformations in the Earth's crust.

Beneath the earth's crust lies the mantle which is heated by the radioactive decay of heavy elements. The mantle is not quite solid and consists of magma which is in a state of semi-perpetual convection. This convection process causes the lithospheric plates to move, albeit slowly. The resulting process is known as plate tectonics.

Plate tectonics might be thought of as the process by which the earth resurfaces itself. Through a process called *spreading ridges* (or seafloor spreading), the earth creates new crust by allowing magma underneath the lithosphere to come to the surface where it cools and solidifies—becoming new crust, and through a process called subduction, excess crust is pushed underground—beneath the rest of the lithosphere—where it comes into contact with magma and melts—rejoining the mantle from which it originally came.

Areas of the crust where new crust is created are called *divergent boundaries*, and areas of the crust where it is brought back into the earth are called *convergent boundaries*. Earthquakes result from the movement of the lithospheric plates, and they often occur near covergent boundaries where parts of the crust are forced into the earth as part of subduction.

Volcanoes result primarily from the melting of subducted crust material. Crust material that is forced into the athenosphere melts, and some portion of the melted material becomes light enough to rise to the surface—giving birth to volcanoes.

Earth's Electromagnet

An electromagnet is a magnet that is created by a current that flows around a soft-iron core. The earth has a soft iron core surrounded by semi-liquid materials from the mantle that move in continuous currents around the core; therefore, the earth is an electromagnet. This is referred to as the dynamo theory of earth's magnetism. The fact that earth is an electromagnet helps with the earth's maintenance of an atmosphere suitable for life.

Atmosphere

The earth is blanketed by an atmosphere consisting of 99% oxygen and nitrogen. The atmosphere has five layers: troposphere, stratosphere, mesosphere, thermosphere, and exosphere; and 75% of the atmosphere's gasses are in the bottom-most layer, the troposphere.

The magnetic field created by mantle's internal motions produces the magnetosphere which protects the earth's atmosphere from the solar wind. It is theorized that the solar wind would strip away earth's atmosphere in a few million years were it not for the earth's electromagnet. And since earth is 4.5 billion years old, earth would not have an atmosphere by now if there were no magnetosphere.

The atmosphere is composed of 78% nitrogen and 21% oxygen. The remaining one percent contains small amounts of other gasses including CO_2 and water vapors. Water vapors and CO_2 allow the earth's atmosphere to catch and hold the sun's energy through a phenomenon called the greenhouse effect. This allows earth's surface to be warm enough to have liquid water and support life.

In addition to storing heat, the atmosphere also protects living organisms by shielding the earth's surface from cosmic rays. Note that the level of protection seems to be adjusted to prevent cosmic rays from destroying all life while allowing for the mutations that have an important role in pushing forward diversity in the biosphere.

Methodology

Like all other scientists, Earth scientists apply the scientific method. They formulate hypotheses after observing events and gathering data about natural phenomena, and then they test those hypotheses. In Earth science, data usually plays a major role in testing and formulating hypotheses.

A driving philosophical force behind contemporary earth science is uniformitarianism. Uniformitarianism says that "ancient geologic features are interpreted by understanding active processes that are readily observed". Simply stated, this means that dramatic and awe-inspiring features of the earth can be explained by the actions of gradual processes operating over long periods of time; for example, a mountain need not be thought of as having been created in a moment, but instead it may be seen as the result of earthquake after earthquake raising the ground in a certain area by small amounts over millions of years.

Partial List of the Major Earth Science Topics

Atmosphere

- Atmospheric chemistry
- Climatology
- Meteorology
    - Hydrometeorology
- Paleoclimatology

Biosphere

- Biogeography
- Paleontology
  - Palynology
  - Micropaleontology
- Geomicrobiology

Hydrosphere

- Hydrology
  - Glaciology
  - Limnology
- Hydrogeology
- Oceanography
  - Chemical oceanography
  - Marine biology
  - Marine geology
  - Paleoceanography
  - Physical oceanography

Lithosphere or Geosphere

- Geology
  - Economic geology
  - Engineering geology
  - Environmental geology
  - Historical geology
    - Glaciology
    - Quaternary geology
  - Planetary geology
  - Sedimentology
  - Stratigraphy
  - Structural geology
- Geochemistry

- Geography
  - Physical geography
- Geomorphology
- Geophysics
  - Geochronology
  - Geodynamics (see also Tectonics)
  - Geomagnetics
  - Gravimetry (also part of Geodesy)
  - Seismology
- Hydrogeology
- Mineralogy
  - Crystallography
  - Gemology
- Petrology
- Volcanology

Pedosphere

- Soil science
  - Edaphology
  - Pedology

Systems

- Environmental science
- Geography
  - Human geography
  - Physical geography
- Gaia hypothesis

Others

- Cartography
- Geoinformatics (GIS)
- Geostatistics
- Geodesy and Surveying

4

Energy Development and Renewable Energy: Select Case Studies

ENERGY DEVELOPMENT

Energy development is the ongoing effort to provide sustainable energy resources through knowledge, skills, and constructions. When harnessing energy from primary energy sources and converting them into more convenient secondary energy forms, such as electrical energy and cleaner fuel, both emissions (reducing pollution) and quality (more efficient use) are important.

Dependence on External Energy Sources

Technologically advanced societies have become increasingly dependent on external energy sources for transportation, the production of many manufactured goods, and the delivery of energy services. This energy allows people, in general, to live under otherwise unfavorable climatic conditions through the use of heating, ventilation, and/or air conditioning. Level of use of external energy sources differs across societies, as do the climate, convenience, traffic congestion, pollution, production, and greenhouse gas emissions of each society.

Increased levels of human comfort generally induce increased dependence on external energy sources, although the application of energy efficiency and conservation approaches allows a certain degree of mitigation of the dependence. Wise energy use therefore embodies the idea of balancing human comfort with reasonable energy consumption levels by researching and implementing effective and sustainable energy harvesting and utilization measures.

Limitations to Energy Development

A key limit to the development of any particular energy source is availability of the underlying resource. Most of the world's main energy sources are based on the consumption of non-renewable resources (petroleum, coal, natural gas, and uranium). While still a small segment of the energy supply, renewable sources such as wind power and solar power are growing rapidly in market share.

Closely linked to energy development are concerns about the possible environmental effects of energy use, such as climate changes. Energy development issues are part of the much debated sustainable development problem.

Primary Energy Sources

Primary energy sources are substances or processes with concentrations of energy at a high enough potential to be feasibly encouraged to convert to lower energy forms under human control for human benefit. Except for nuclear fuels, tidal energy and geothermal energy, all terrestrial energy sources are from current solar insolation or from fossil remains of plant and animal life that relied directly and indirectly upon sunlight, respectively. And ultimately, solar energy itself is the result of the Sun's nuclear fusion. Geothermal power from hot, hardened rock above the magma of the earth's core is the result of the accumulation of radioactive materials during the formation of Earth which was the byproduct of a previous supernova event.

Fossil Fuels

Fossil fuels, in terms of energy, involve the burning of coal or hydrocarbon fuels, which are the remains of the decomposition of plants and animals. Steam power plant combustion heats water to create steam, which turns a turbine, which, in turn, generates electricity, waste heat, and pollution. There are three main types of fossil fuels: coal, petroleum, and natural gas. Another fossil fuel, liquefied petroleum gas (LPG), is principally derived from the production of natural gas.

Pros

- Because it is based on the simple process of combustion, the burning of fossil fuels can generate large amounts of electricity with a small amount of fuel. Gas-fired power plants are more efficient than coal fired power plants.
- Fossil fuels such as coal are readily available and are currently plentiful. Excluding external costs, coal is less expensive than most other sources of energy because there are large deposits of coal in the world
- The technology already exists for the use of fossil fuels, though oil and natural gas are approaching peak production and will require a transition to other fuels and/or other measures.
- Commonly used fossil fuels in liquid form such as light crude oil, gasoline, and LPG are easy to distribute.

- LPG can be transported, stored and used virtually anywhere. It does not require a fixed network and will not deteriorate over time. As a result, it is particularly useful in regions which are not connected to fixed energy networks.
- LPG is clean burning and has lower greenhouse gas emissions than any other fossil fuel when measured on a total fuel cycle. In fact, by 2010, all buses and taxis in the Southern Chinese city of Guangzhou will be LP Gas fueled. The city will host the 2010 Asian games and has taken the step in a bid to reduce air pollution in advance of the games. LPG is also non-toxic and will not contaminate soil or aquifers in the event of a leak.
- LPG can be accessible to everyone everywhere today without major infrastructure investment. There are enough reserves to last many decades.
- LPG can be up to 5 times more efficient than traditional fuels, resulting in less energy wastage and better use of our planet's resources.

Cons

- The combustion of fossil fuels leads to the release of pollution into the atmosphere. According to the Union of Concerned Scientists, a typical coal plant produces in one year:
  - 3,700,000 tons of carbon dioxide (CO_2), the primary human cause of global warming.
  - 10,000 tons of sulfur dioxide (SO_2), the leading cause of acid rain.
  - 500 tons of small airborne particles, which result in chronic bronchitis, aggravated asthma, and premature death, in addition to haze-obstructed visibility.
  - 10,200 tons of nitrogen oxides (NO_x), leading to formation of ozone (smog) which inflames the lungs, burning lung tissue making people more susceptible to respiratory illness.
  - 720 tons of carbon monoxide (CO), resulting in headaches and additional stress on people with heart disease.
  - 220 tons of hydrocarbons, volatile organic compounds (VOC), which form ozone.
  - 170 pounds of mercury, where just 1/70th of a teaspoon deposited on a 25-acre lake can make the fish unsafe to eat.
  - 225 pounds of arsenic, which will cause cancer in one out of 100 people who drink water containing 50 parts per billion.

- 114 pounds of lead, 4 pounds of cadmium, other toxic heavy metals, and trace amounts of uranium.

- Dependence on fossil fuels from volatile regions or countries creates energy security risks for dependent countries. Oil dependence in particular has led to monopolization, war, and socio-political instability.
- They are considered non-renewable resources, which will eventually decline in production and become exhausted, with dire consequences to societies that remain highly dependent on them. Fossil fuels are actually slowly forming continuously, but we are using them up at a rate approximately 100,000 times faster than they are formed.
- Extracting fossil fuels is becoming more difficult as we consume the most accessible fuel deposits. Extraction of fossil fuels is becoming more expensive and more dangerous as mines get deeper and oil rigs go further out to sea.
- Extraction of fossil fuels can result in extensive environmental degradation, such as the strip mining and mountaintop removal of coal.
- The drilling and transportation of petroleum can result in accidents that result in the despoilation of hundreds of kilometers of beaches and the death or elimination of many forms of wildlife in the area
- Safety measures are necessary in order to use LPG without incident.
- The storage of these fuels can result in accidents with explosions and poisoning of the atmosphere and groundwater.

Biomass, Biofuels, and Vegetable Oil

Biomass production involves using garbage or other renewable resources such as corn or other vegetation, to generate electricity. When garbage decomposes the methane produced is captured in pipes and later burned to produce electricity. Vegetation and wood can be burned directly, like fossil fuels, to generate energy, or processed to form alcohols.

Vegetable oil is generated from sunlight and CO2 by plants. It is safer to use and store than gasoline or diesel as it has a higher flash point. Straight vegetable oil works in diesel engines if it is heated first. Vegetable oil can also be transesterified to make biodiesel which burns like normal diesel.

Pros

- Biomass production can be used to burn organic waste products resulting from agriculture. This type of recycling encourages the philosophy that nothing on this Earth should be wasted. The result is less demand on the

Earth's resources, and a higher carrying capacity for Earth because non-renewable fossil fuels are not consumed.

- Biomass is abundant on Earth and is generally renewable. In theory, we will never run out of organic waste products as fuel, because we are continuously producing them. In addition, biomass is found throughout the world, a fact that should alleviate energy pressures in third world nations.
- When methods of biomass production other than direct combustion of plant mass, such as fermentation and pyrolysis, are used, there is little effect on the environment. Alcohols and other fuels produced by these alternative methods are clean burning and are feasible replacements to fossil fuels.
- Since CO2 is first taken out of the atmosphere to make the vegetable oil and then put back after it is burned in the engine, there is no net increase in CO2. So vegetable oil does not contribute to the problem of greenhouse gas.
- Vegetable oil has a higher flash point and is safer than most fossil fuels.
- Transitioning to vegetable oil could be relatively easy as biodiesel works where diesel works, and straight vegetable oil takes relatively minor modifications.
- The World already produces more than 100 billion gallons a year for food industry, so we have experience making it.
- Algaculture has the potential to produce far more vegetable oil per acre than current plants.
- Infrastructure for biodiesel around the World is significant and growing.

Cons

- Direct combustion without emissions filtering generally leads to air pollution similar to that from fossil fuels.
- Producing liquid fuels from biomass is generally less cost effective than from petroleum, since the production of biomass and its subsequent conversion to alcohols is particularly expensive.
- Some researchers claim that, when biomass crops are the product of intensive farming, ethanol fuel production results in a net loss of energy after one accounts for the fuel costs of fertilizer production, farm equipment, and the distillation process.
- Direct competition with land use for food production.
- Current production methods would require enormous amounts of land to replace all gasoline and diesel. With current technology, it is unfeasible for biofuels to replace the demand for petroleum.

- Growth in vegetable oil production is already resulting in deforestation.
- Converting forest land to vegetable oil production can result in a net increase in CO2.
- Demand for vegetable oil used as a fuel may drive up prices of vegetable oils in the food industry
- Costs to modify existing engines may outweigh fuel cost savings

Hydroelectric Energy

In hydro energy, the gravitational descent of a river is compressed from a long run to a single location with a dam or a flume. This creates a location where concentrated pressure and flow can be used to turn turbines or water wheels, which drive a mechanical mill or an electric generator.

Pros

- Hydroelectric power stations can promptly increase to full capacity, unlike other types of power stations. This is because water can be accumulated above the dam and released to coincide with peaks in demand.
- Electricity can be generated constantly, so long as sufficient water is available.
- Hydroelectric power produces no primary waste or pollution.
- Hydropower is a renewable resource.
- Hydroelectricity assists in securing a country's access to energy supplies.

Cons

- The construction of a dam can have a serious environmental impact on the surrounding areas. The amount and the quality of water downstream can be affected, which affects plant life both aquatic, and land-based. Because a river valley is being flooded, the delicate local habitat of many species are destroyed, while people living nearby may have to relocate their homes.
- Hydroelectricity can only be used in areas where there is a sufficient supply of water.
- Flooding submerges large forests (if they have not been harvested). The resulting anaerobic decomposition of the carboniferous materials releases methane, a greenhouse gas.
- Dams can contain huge amounts of water. As with every energy storage system, failure of containment can lead to catastrophic results, e.g. flooding.
- Hydroelectric plants rarely can be erected near load centres, requiring large transmission lines.

Nuclear Energy

Nuclear power stations use nuclear fission to generate energy by the reaction of uranium-235 inside a nuclear reactor. The reactor uses uranium rods, the atoms of which are split in the process of fission, releasing a large amount of energy. The process continues as a chain reaction with other nuclei. The heat released heats water to create steam, which spins a turbine generator, producing electricity. A relatively small number of nuclear power plants (about 50) has the potential to supply the entire U.S.

Depending on the type of fission fuel considered, estimates for existing supply at known usage rates varies from several decades for the currently popular Uranium-235 to thousands of years for uranium-238. At the present use rate, there are (as of 2007) about 70 years left of known uranium-235 reserves economically recoverable at an uranium price of US$ 130/kg. The nuclear industry argue that the cost of fuel is a minor cost factor for fission power, more expensive, more difficult to extract sources of uranium could be used in the future, such as lower-grade ores, and if prices increased enough, from sources such as granite and seawater. Increasing the price of uranium would have little effect on the overall cost of nuclear power; a doubling in the cost of natural uranium would increase the total cost of nuclear power by 5 percent. On the other hand, if the price of natural gas was doubled, the cost of gas-fired power would increase by about 60 percent.

Opponents on the other hand argue that the correlation between price and production is not linear, but as the ores' concentration are becoming smaller, the difficulty (energy and resource consumption are increasing, while the yields are decreasing) of extraction is rising very fast, and that the assertion that a hear price will yield more uranium is overly optimistic; for example a ruffle estimate predicts that the extraction of uranium from granite will consume at least 70 times more energy than what it will produce in a reactor. Seawater seems to be equally dubious as a source. As a consequence an eventual doubling in the price of uranium will give a marginal increase in the volumes that are being produced.

Another alternative would be to use thorium as fission fuel. Thorium is three times more abundant in Earth's crust than uranium, and much more of the thorium can be used (or, more precisely, converted into Uranium-233 and then used).

Current light water reactors burn the nuclear fuel poorly, leading to energy waste. Nuclear reprocessing or burning the fuel better using different reactor designs would reduce the amount of waste material generated and allow better use of the available resources. As opposed to current light water reactors which use uranium-235 (0.7 percent of all natural uranium), fast breeder reactors convert the more abundant uranium-238 (99.3 percent of all natural uranium) into plutonium for fuel. It has been estimated that there is anywhere from 10,000 to five billion years worth of Uranium-238 for use in these power plants . Breeder technology has been used in several reactors. However, the breeder reactors at Dounreay in Scotland, Monju in Japan and the Superphénix at Creys-Malville in France, in particular, have all had difficulties and were not economically competitive and have been decommissioned. The People's Republic of China intends to build breeders.

The possibility of nuclear meltdowns and other reactor accidents, such as the Three Mile Island accident and the Chernobyl disaster, have caused much public fear. Research is being done to lessen the known problems of current reactor technology by developing automated and passively-safe reactors. Historically, however, coal and hydropower power generation have both been the cause of more deaths per energy unit produced than nuclear power generation. Various kinds of energy infrastructure might be attacked by terrorists, including nuclear power plants, hydropower plants, and liquified natural gas tankers. Nuclear proliferation is the spread from nation to nation of nuclear technology, including nuclear power plants but especially nuclear weapons. New technology like SSTAR ("small, sealed, transportable, autonomous reactor") may lessen this risk.

The long-term radioactive waste storage problems of nuclear power have not been fully solved. Several countries have considered using underground repositories. Nuclear waste takes up little space compared to wastes from the chemical industry which remain toxic indefinitely. Spent fuel rods are now stored in concrete casks close to the nuclear reactors. The amounts of waste could be reduced in several ways. Both nuclear reprocessing and fast breeder reactors could reduce the amounts of waste. Subcritical reactors or fusion reactors could greatly reduce the time the waste has to be stored. Subcritical reactors may also be able to do the same to already existing waste. The only way of dealing this wastes today is by geological storage.

The economics of nuclear power is not simple to evaluate, because of high capital costs for building and very low fuel costs. Comparison with other power generation methods is strongly dependent on assumptions about construction timescales and capital financing for nuclear plants. See Economics of new nuclear power plants.

Depending on the source different energy return on energy investment (EROI) are claimed. Advocates (using life cycle analysis) argue that it takes 4-5 months of energy production from the nuclear plant to fully pay back the initial energy investment. Opponents claim that it depends on the grades of the ores the fuel came from, so a full payback can vary from 10 to 18 years, and that the advocates' claim was based on the assumption of high grade ores (the yields are getting worst, as the ores are leaner, for less than 0.02% ores,the yield is less then 50%).

Advocates also claim that it is possible to relatively rapidly increase the number of plants. Typical new reactor designs have a construction time of three to four years. In 1983, 43 plants were being built, before an unexpected fall in fossil fuel prices stopped most new construction. Developing countries like India and China are rapidly increasing their nuclear energy use. However, a Council on Foreign Relations report on nuclear energy argues that a rapid expansion of nuclear power may create shortages in building materials such as reactor-quality concrete and steel, skilled workers and engineers, and safety controls by skilled inspectors. This would drive up current prices.

Pros

- The energy content of a kilogram of uranium or thorium, if spent nuclear fuel is reprocessed and fully utilized, is equivalent to about 3.5 million kilograms of coal.
- The cost of making nuclear power, with current legislation, is about the same as making coal power, which is considered very inexpensive (see Economics of new nuclear power plants). If a carbon tax is applied, nuclear does not have to pay anything because nuclear does not emit toxic gases such as CO_2, NO, CO, SO_2, arsenic, etc. that are emitted by coal power plants.
- Nuclear power plants are guarded with the nuclear reactor inside a reinforced containment building, and thus are relatively impervious to terrorist attack or adverse weather conditions (see Nuclear safety in the U.S.).
- Because of the fear of a nuclear disaster, nuclear safety has become a major issue.
- Nuclear power does not produce any primary air pollution or release carbon dioxide and sulfur dioxide into the atmosphere. Therefore, it contributes only a small amount to global warming or acid rain.
- Coal mining is the second most dangerous occupation in the United States. Nuclear energy is much safer per capita than coal derived energy.
- For the same amount of electricity, the life cycle emissions of nuclear is about 4% of coal power. Depending on the report, hydro, wind, and geothermal are sometimes ranked lower, while wind and hydro are sometimes ranked higher (by life cycle emissions).
- According to a Stanford study, fast breeder reactors have the potential to power humans on earth for billions of years, making it sustainable.

Cons

- The operation of an uncontained nuclear reactor near human settlements can be catastrophic, as shown by the Chernobyl disaster in the Ukraine (former USSR), where large areas of land were affected by radioactive contamination.

- Waste produced from nuclear fission of uranium is both poisonous and highly radioactive, requiring maintenance and monitoring at the storage sites. Moreover, the long-term disposal of the long-lived nuclear waste causes serious problems, since (unless the spent fuel is reprocessed) it takes from one to three thousand years for the spent fuel to come back to the natural radioactivity of the uranium ore body that was mined to produce it.
- There can be connections between nuclear power and nuclear weapon proliferation, since many reactor designs require large-scale uranium enrichment facilities. While civilian nuclear facilities are normally overseen internationally by the IAEA, several countries with such facilities refuse oversight
- Large capital cost. Building a nuclear power plant requires a huge investment and the costs of safe disassembling (called decommissioning) after it reaches end of usable life must be factored into the full lifecycle budget (see Economics of new nuclear power plants
- Nuclear fuels are a non-renewable energy source, with unknown high concentration ore reserves. There is a large amount of trace concentration nuclear material in seawater and most rocks; however, extraction from these is not currently economically competitive.
- The limited liability for the owner of a nuclear power plant in case of a nuclear accident differs per nation while nuclear installations are sometimes built close to national borders.
- Waste heat disposal becomes an issue at high ambient temperature thus at a time of peak demand the reactor may need to be shut down or have reduced output

Fusion Power

Fusion power could solve many of the problems of fission power (the technology mentioned above) but, despite research having started in the 1950s, no commercial fusion reactor is expected before 2050 . Many technical problems remain unsolved. Proposed fusion reactors commonly use deuterium, an isotope of hydrogen, as fuel and in most current designs also lithium. Assuming a fusion energy output equal to the current global output and that this does not increase in the future, then the known current lithium reserves would last 3000 years, lithium from sea water would last 60 million years, and a more complicated fusion process using only deuterium from sea water would have fuel for 150 billion years.

Wind Power

This type of energy harnesses the power of the wind to propel the blades of wind turbines. These turbines cause the rotation of magnets, which creates electricity. Wind towers are usually built together on wind farms.

Pros

- Wind power produces no water or air pollution that can contaminate the environment, because there are no chemical processes involved in wind power generation. Hence, there are no waste by-products, such as carbon dioxide.
- Power from the wind does not contribute to global warming because it does not generate greenhouse gases.
- Wind generation is a renewable source of energy, which means that we will never run out of it.
- Wind towers can be beneficial for people living permanently, or temporarily, in remote areas. It may be difficult to transport electricity through wires from a power plant to a far-away location and thus, wind towers can be set up at the remote setting.
- Farming and grazing can still take place on land occupied by wind turbines.
- Those utilizing wind power in a grid-tie configuration will have backup power in the event of a grid outage.
- Due to the ability of wind turbines to coexist within agricultural fields, siting costs are frequently low.

Cons

- Wind is unpredictable, therefore wind power is not predictably available. When the wind speed decreases less electricity is generated.
- Wind farms may be challenged in communities that consider them an eyesore or view obstructor.
- Wind farms, depending on the location and type of turbine, may negatively affect bird migration patterns and may pose a danger to the birds themselves. Newer, larger wind turbines have slower moving blades which are visible to birds.

Solar Power

Solar power involves using solar cells to convert sunlight into electricity, using sunlight hitting solar thermal panels to convert sunlight to heat water or air, using sunlight hitting a parabolic mirror to heat water (producing steam), or using sunlight entering windows for passive solar heating of a building. It would be advantageous to place solar panels in the regions of highest solar radiation. In the Phoenix, Arizona area, for example, the average annual solar radiation is 5.7 kWh/m2/day, or 2080.5 kWh/m2/year. Electricity demand in the continental U.S. is $3.7*10^{12}$ kW·h per year. Thus, at 100% efficiency, an area of 1.8x10^9 sq. m (around 700 sq miles) would need to be covered with solar panels to replace all current electricity production in the US with solar power, and at 20% efficiency, an area of approximately 3500 sq miles (3% of Arizona's land area). The average solar radiation in the United States is 4.8 kwh/m2/day, but reaches 8-9 kWh/m2/day in parts of Southwest.

The monetary cost, assuming $500/meter², would be about $5-10 trillion dollars.

Pros

- Solar power imparts no fuel costs.
- Solar power is a renewable resource. As long as the Sun exists, its energy will reach Earth.
- Solar power generation releases no water or air pollution, because there is no combustion of fuels.
- In sunny countries, solar power can be used in remote locations, like a wind turbine. This way, isolated places can receive electricity, when there is no way to connect to the power lines from a plant.
- Solar energy can be used very efficiently for heating (solar ovens, solar water and home heaters) and daylighting.
- Requires no fuel.
- Coincidently, solar energy is abundant in regions that have relatively largest number of people living off grid - in developing regions of Africa, Indian subcontinent and Latin America. Hence cheap solar, when availabile, opens the opportunity to enhance global electricity access considerably, and possibly in a relatively short time period.

Cons

- Solar electricity is expensive compared to grid electricity.
- Solar heat and electricity are not available at night and may be unavailable due to weather conditions; therefore, a storage or complementary power system is required for most applications.
- Limited power density: Average daily insolation in the contiguous U.S. is 3-7 kW·h/m² [34][35] (see picture)

- Solar cells produce DC which must be converted to AC (using a grid tie inverter) when used in currently existing distribution grids. This incurs an energy loss of 4-12%.
- A photovoltaic power station is expensive to build, and the energy payback time - the time necessary for producing the same amount of energy as needed for building the power device - for photovoltaic cells is about 1-5 years, depending primarily on location.
- Solar panels collect dust and require cleaning. Dust on the panels significantly reduces the transfer of energy from solar radiation to electric current.

Geothermal Energy

Geothermal energy harnesses the heat energy present underneath the Earth. The hot rocks heat water to produce steam. When holes are drilled in the region, the steam that shoots up is purified and is used to drive turbines, which power electric generators. When the water temperature is below the boiling point of water a binary system is used. A low boiling point liquid is used to drive a turbine and generator in a closed system similar to a refrigeration unit running in reverse.

Pros

- Geothermal energy does not produce air or water pollution if performed correctly.
- Geothermal power plants run continuously day and night with an uptime typically exceeding 95%.
- Once a geothermal power station is implemented, the energy produced from the station is practically free. A small amount of energy is required in order to run a pump, although this pump can be powered by excess energy generated at the plant.
- Geothermal power stations are relatively small, and have a lesser impact on the environment than tidal or hydroelectric plants. Because geothermal technology does not rely on large bodies of water, but rather, small, but powerful jets of water, like geysers, large generating stations can be avoided without losing functionality.

Cons

- Geothermal energy extraction is only practical in certain areas of the world, usually volcanic, where the heated rock is sufficiently close to the surface such as to be reached by current drilling technology . Enhanced geothermal technology uses deeper drilling and water injection to generate geothermal power in areas where the earth's crust is thicker.

- Drilling holes underground may release hazardous gases and minerals from deep inside the Earth. It can be problematic to dispose of these subsidiary products in a safe manner.

Energy Transportation

While new sources of energy are only rarely discovered or made possible by new technology, distribution technology continually evolves. The use of fuel cells in cars, for example, is an anticipated delivery technology. This section presents some of the more common delivery technologies that have been important to historic energy development. They all rely in some way on the energy sources listed in the previous section.

Fuels

Shipping is a flexible delivery technology that is used in the whole range of energy development regimes from primitive to highly advanced. Currently, coal, petroleum and their derivatives are delivered by shipping via boat, rail, or road. Petroleum and natural gas may also be delivered via pipeline and coal via a Slurry pipeline. Refined hydrocarbon fuels such as gasoline and LPG may also be delivered via aircraft. Natural gas pipelines must maintain a certain minimum pressure to function correctly

Electric Grids

Electricity grids are the networks used to transmit and distribute power from production source to end user, when the two may be hundreds of kilometres away. Sources include electrical generation plants such as a nuclear reactor, coal burning power plant, etc. A combination of sub-stations, transformers, towers, cables, and piping are used to maintain a constant flow of electricity.

Grids may suffer from transient blackouts and brownouts, often due to weather damage. During certain extreme space weather events solar wind can interfere with transmissions.

Grids also have a predefined carrying capacity or load that cannot safely be exceeded. When power requirements exceed what's available, failures are inevitable. To prevent problems, power is then rationed.

Industrialised countries such as Canada, the US, and Australia are among the highest per capita consumers of electricity in the world, which is possible thanks to a widespread electrical distribution network. The US grid is one of the most advanced, although infrastructure maintenance is becoming a problem. The electrical power industry is one of the most heavily subsidized.

CurrentEnergy provides a realtime overview of the electricity supply and demand for California, Texas, and the Northeast of the US. African countries with small scale electrical grids have a correspondingly low annual per capita usage of electricity. One of the most powerful power grids in the world supplies power to the state of Queensland, Australia.

Energy Storage

Methods of energy storage have been developed, which transform electrical energy into forms of potential energy. A method of energy storage may be chosen based on stability, ease of transport, ease of energy release, or ease of converting free energy from the natural form to the stable form.

Compressed Air Vehicles

An Indian company is planning to release a compressed air powered car in 2008.

Battery-Powered Vehicles

Batteries are used to store energy in a chemical form. As an alternative energy, batteries can be used to store energy in battery electric vehicles. Battery electric vehicles can be charged from the grid when the vehicle is not in use. Because the energy is derived from electricity, battery electric vehicles make it possible to use other forms of alternative energy such as wind, solar, geothermal, nuclear, or hydroelectric.

Pros

- Produces zero emissions to help counteract the effects of global warming.
- Batteries are a mature technology, no new expensive research and development is needed to implement technology.
- Current lead acid battery technology offers 50+ miles range on one charge.
- The Tesla Roadster has a 200 mile range on one charge.
- Batteries make it possible for stationary alternative energy generation such as solar, wind, hydroelectric, or nuclear
- Electric motors are 90% efficient compared to about 20% efficiency of an internal combustion engine.
- No new major infrastructure is needed to charge battery electric vehicles.
- Battery electric vehicles have fewer moving parts than internal combustion engines, thus improving the reliability of the vehicle.
- Battery electric vehicles are quiet compared to internal combustion engines.

- Multiple electric vehicles sold out including the General Motors EV1 and the Tesla Roadster proving the demand for battery electric vehicles.
- Operation of a battery electric vehicle is approximately 2 to 4 cents per mile. About a sixth the price of operating a gasoline vehicle.
- The use of Battery Electric Vehicles eliminates the dependency on foreign oil.

Cons

- Current battery technology is expensive.
- Battery electric vehicles have a relative short range compared to internal combustion engine vehicles.

Hydrogen Economy

Hydrogen can be manufactured at roughly 77 percent thermal efficiency by the method of steam reforming of natural gas. When manufactured by this method it is a derivative fuel like gasoline; when produced by electrolysis of water, it is a form of chemical energy storage as are storage batteries, though hydrogen is the more versatile storage mode since there are two options for its conversion to useful work: (1) a fuel cell can convert the chemicals hydrogen and oxygen into water, and in the process, produce electricity, or (2) hydrogen can be burned (less efficiently than in a fuel cell) in an internal combustion engine.

Pros

- Hydrogen is colorless, odorless and entirely non-polluting, yielding pure water vapor (with minimal NOx) as exhaust when combusted in air. This eliminates the direct production of exhaust gases that lead to smog, and carbon dioxide emissions that enhance the effect of global warming.
- Hydrogen is the lightest chemical element and has the best energy-to-weight ratio of any fuel (not counting tank mass).
- Hydrogen can be produced anywhere; it can be produced domestically from the decomposition of water. Hydrogen can be produced from domestic sources and the price can be established within the country.
- Electrolysis combined with fuel-cell regeneration is more than 50% efficient.

Cons

- Other than some volcanic emanations, hydrogen does not exist in its pure form in the environment, because it reacts so strongly with oxygen and other elements.

- It is impossible to obtain hydrogen gas without expending energy in the process. There are three ways to manufacture hydrogen;
  - By breaking down hydrocarbons — mainly methane. If oil or gases are used to provide this energy, fossil fuels are consumed, forming pollution and nullifying the value of using a fuel cell. It would be more efficient to use fossil fuel directly.
  - By electrolysis from water — The process of splitting water into oxygen and hydrogen using electrolysis consumes large amounts of energy. It has been calculated that it takes 1.4 joules of electricity to produce 1 joule of hydrogen (Pimentel, 2002).
  - By reacting water with a metal such as sodium, potassium, or boron. Chemical by-products would be sodium oxide, potassium oxide, and boron oxide. Processes exist which could recycle these elements back into their metal form for re-use with additional energy input, further eroding the energy return on energy invested.
- There is currently modest fixed infastructure for distribution of hydrogen that is centrally produced amounting to several hundred kilometers of pipeline. An alternative would be transmission of electricity over the existing electrical network to small-scale electrolyzers to support the widespread use of hydrogen as a fuel.
- Hydrogen is difficult to handle, store, and transport. It requires heavy, cumbersome tanks when stored as a gas, and complex insulating bottles if stored as a cryogenic liquid. If it is needed at a moderate temperature and pressure, a metal hydride absorber may be needed. The transportation of hydrogen is also a problem because hydrogen leaks effortlessly from containers.
- Some current fuel cell designs, such as proton exchange membrane fuel cells, use platinum as a catalyst. Widescale deployment of such fuel cells could place a strain on available platinum resources. [43] Reducing the platinum loading, per fuel cell stack, is the focus of R&D.
- Electricity transmission and battery electric vehicles are far more efficient for storage, transmission and use of energy for transportation, neglecting the energy conversion at the electric power plant. As with distributed production of hydrogen via electrolysis, battery electric vehicles could utilize the existing electricity grid until widespread use dictated an expansion of the grid.

Energy Storage Types

Chemical

Some natural forms of energy are found in stable chemical compounds such as fossil fuels. Most systems of chemical energy storage result from biological activity, which store energy in chemical bonds. Man-made forms of chemical energy storage include hydrogen fuel, batteries and explosives such as cordite and dynamite.

Gravitational

Dams can be used to store energy, by using excess energy to pump water into the reservoir. When electrical energy is required, the process is reversed. The water then turns a turbine, generating electricity. Hydroelectric power is currently an important part of the world's energy supply, generating one-fifth of the world's electricity.:

Electrical Capacitance

Electrical energy may be stored in capacitors. Capacitors are often used to produce high intensity releases of energy (such as a camera's flash).

Mechanical

- *Pressure:* Energy may also be stored pressurized gases or alternatively in a vacuum. Compressed air, for example, may be used to operate vehicles and power tools. Large scale compressed air energy storage facilities are used to smooth out demands on electricity generation by providing energy during peak hours and storing energy during off-peak hours. Such systems save on expensive generating capacity since it only needs to meet average consumption rather than peak consumption.
- *Flywheels and springs:* Energy can also be stored in mechanical systems such as springs or flywheels. Flywheel energy storage is currently being used for uninterruptible power supplies.

Future Energy Development

Extrapolations from current knowledge to the future offer a choice of energy futures. Some predictions parallel the Malthusian catastrophe hypothesis. Numerous are complex models based scenarios as pioneered by *Limits to Growth*. Modeling approaches offer ways to analyze diverse strategies, and hopefully find a road to rapid and sustainable development of humanity. Short term energy crises are also a concern of energy development. Some extrapolations lack plausibility, particularly when they predict a continual increase in oil consumption.

Existing technologies for new energy sources, such as renewable energy technologies, particularly wind power and solar power, are promising. Nuclear fission is also promoted, and each need sustained research and development, including consideration of possible harmful side effects. Jacques Cousteau spoke of using the salinization of water at river estuaries as an energy source, which would not have any consequences for a million years, and then stopped to point out that since we are going to be on the planet for a billion years we had to be looking that far into the future. Nuclear fusion and artificial photosynthesis are other energy technologies being researched and developed.

It should be noted that between 1950 and 1984, as the Green Revolution transformed agriculture around the globe, world grain production increased by 250%. The energy for the Green Revolution was provided by fossil fuels in the form of fertilizers (natural gas), pesticides (oil), and hydrocarbon fueled irrigation. The peaking of world hydrocarbon production (Peak oil) may test Malthus critics.

EMERGY

The term Emergy was originally coined by David M. Scienceman in collaboration with the late Howard T. Odum. H.T.Odum used 'emergy' to mean both *sequestered* energy and *emergent property of energy use*. However Scienceman also used the term 'emergy' to refer to the concept of *energy memory*, a concept which motivated Scienceman and B.M. El-Youssef (1993) to propose the addition of a new dimension to physics, [P] for "past". Some researchers maintain that it can be expressed as a scientific unit which is called the "emjoule", a contraction of "*em*ergy joule".

Historical Development

As a word, emergy is a simple contraction of the term "embodied energy". The need for the new word of "emergy" arose apparently because of an important difference in the way the two related disciplines of systems ecology and energy analysis were using the term "embodied energy". As H.T.Odum observed "There is more than one type of embodied energy". Various authors have struggled to clarify their usage, and ambiguity seems to continue in the literature to this day.

Prior to 1986, both systems ecologists and energy analysts used "embodied energy" to refer to the sum over time of all energy of one type required to generate a flow of energy. Energy embodied in water was also defined in this way as the energy required directly and indirectly to generate the flow in processes of the biosphere or a typical desalination plant for example.

However unlike energy analysts, systems ecologists were, and continue to be also interested in the relationship of structure and function of ecological systems, together with the biophysical properties that afford plants the ability to accumulate and transform radiant energy into a structural form usable by other organisms, Scienceman observed this difference noting that the systems ecologist H.T. Odum had introduced an additional factor into the definition of embodied energy:

"Odum ... introduces the maximum power principle into his definition: "The concept here retains the meaning that the embodied energy is what is required to do the work (at maximum power). Byproducts of a work transformation have the same embodied energy because they couldn't be generated with less""

The term "embodied energy" was therefore also used by systems ecologists to describe the energy that had been used in, and accumulated into structure-development and which could be fed back into the system to draw in more energy. This structural-cybernetic aspect of ecosystem research was apparently not included in the energy-analytic definition of "embodied energy". However the fact that both systems ecologists and energy analysts used the same term but with different levels of content seems to have led to considerable confusion.

Enter Scienceman

It was within this context of ambiguity that Scienceman proposed the emergy nomenclature as a means of simplifying scientific discussions. (But also apparently to unify science). Scienceman proposed that the term "emergy" be used to refer the ecosystem-derived concept of embodied energy to make it distinct from the energy-analytic conception. Ten years after proposing the emergy nomenclature in 1987, D.M.Scienceman wrote that his suggestion came as a consequence of studying H.T.Odum's book *Systems Ecology* (later published as *Ecological and General Systems* 1994) for a period of about 2 years. With a background in, "mathematics, practical and theoretical physics (including Einstein's unified field theory) and research in chemical engineering", Scienceman had never before heard of the 'embodied energy' concept, and found Odum's use of it opaque:

> "Owing to great confusion in the scientific literature, and in order to clearly distinguish the theories of H.T.Odum, new concepts - energy memory, emergy, transformity, informity, empower, emtropy, emformation, emtelligence, emprice, emdollars, emtrons and soals - are introduced plus their appropriate units."

With this new terminology, Scienceman sought to clarify two issues: 1/ the combining of energies of different forms; 2/ the process of embodying energies of different forms, or using them up. Scienceman (1997) noticed that the efficiency ratios used in engineering thermodynamics to quantify the transformations of energies from one form into another form were the same kind of ratios that H.T.Odum referred to as energy quality ratios. However in 1986 the National Bureau of Standards did not include any symbol for the notion of 'energy quality', a concept which is pivotal for understanding emergy, and later referred to by the term "energy transformation ratio", or "transformity". Scienceman and H.T.Odum subsequently collaborated on a linguistic project of simplifying and unifying the scientific lexicon by introducing and clarifying new terms. As H.T.Odum later noted,

> "In 1983 our concept of embodied energy (the available energy of one kind previously used up directly and indirectly in transformations to make a product or service) was given the name "EMERGY" and its unit names the "emjoule" or "emcalorie." What is called an "energy transformation ratio" in this chapter was renamed the "transformity" with the unit "emjoule per joule" (not a dimensionless ratio)"

H.T. Odum also used the word 'enmergy' briefly, before using 'emergy' as a standard. Because it can be confused with the word 'energy', some authors use the 'eMergy', and 'EMERGY' notations to emphasize the difference.

Emergy: A Systems Concept

According to the Emergy Systems school, like the concept of sustainability, "Emergy is a systems concept that is context driven, and cannot be fully understood or utilized outside of systems context." This school has therefore promoted emergy as a concept that is useful for establishing the metric for a rigorous and quantitative sustainability index.

The USEPA recently produced an emergy analysis of West Virginia.

Aims of emergy models: integration of human economic and ecological systems

The aim of an emergy model is to quantify the value (see below) of both energy and material resources to humans and non-human systems (e.g., the photosynthetic activity driven by the solar radiation, the dilution of pollutants by the wind, etc.) within a common framework. Hence services provided by the environment which are free and outside the monied economy are included in an emergy model, but may be omitted from an economic model because they are non-monied flows.

This framework attempts to give additional factors to the measurement of value other than monetary cost and consumer's willingness to pay. Non-emergy approaches to the evaluation of ecological, sociopolitical-economic, and industrial processes most often evaluate only non-renewable resources, depending on what human technologies are able to extract from them (user-side quality). Evaluation methodologies seldom have an accounting procedure for the value of human labor, societal services and information. That is, for those flows which carry negligible energy or money, but are supported by a huge indirect flow of resources and perform high quality control, innovation and maintenance actions (contributions to Wikipedia could be considered a good example of such). Emergy evaluation aims to include all of this, perhaps not perfectly, but in a way to help us understand that there is a huge network of supporting energies necessary to support any particular economic activity in human societies and ecological activity in non-human communities. Laganisa and Debeljakb write that

The emergy synthesis method was introduced from Odum in the 1980s ... with the aim of taking into account the different quality of driving forces supporting a process and allowing their comparison on the same basis. It attempts to solve the problem of multi-quality inputs by transforming them to an equivalent of energy of a single quality, which is usually solar energy"

S.E. Jorgensen, S.N.Nielsen and H.Mejer write, "Emergy calculations have the same aim as exergy: to capture the energy hidden in the organization and construction of living organisms."

Method: Draw a System Boundary Diagram and Account for All Inputs and Outputs

Often considered the first step in construction of a quantitative emergy synthesis simulation model is to define the system of interest. This is usually done using the Energy Systems Language to diagrammatically define a systems boundary. This diagram should illustrate all of the flows of emergy and energy into and out of the system of interest, as well as the storages within the system. In addition if system function is influenced by economics, then money flows and stores are also tracked, and again if information is used in system control, this is also diagrammed. The energy, emergy or mass value of these flows is then determined, and placed into an emergy synthesis spreadsheet. These values are then multiplied by either a coefficient to derive a solar emjoule value per joule of available energy in the system of the component (known as a transformity), the solar emjoule value per gram of flow (a specific emergy), or the emergy per dollar value (emdollar).

The purpose of these machinations being to place all components of the system on a common basis, such as the solar energy required to create it. Once this is accomplished it is hoped that the results can be useful to making decisions about public policy, development, trade and engineering that make the overall system more sustainable than it would have been using a non-renewable energy focused evaluation methodology. However in practice there are very few real practical examples in either law, policy making, management accounting, or ecological engineering.

Definition in Words

Emergy can be defined as the total solar equivalent available energy of one form that was used up directly and indirectly in the work of making a product or service.

Emergy expresses the cost of a process or a product in solar energy equivalents.

The basic idea is that solar energy is our ultimate energy source and by expressing the value of products in emergy units, it becomes possible to compare apples and pears. (S.E.Jorgensen 2001, p. 61)

H.T. Odum said that the notion of embodied exergy could be used to evaluate structure (1994, p. 266), and Chen (in press) goes further to define embodied exergy as emergy. For Shu-Li Huang and Chia-Wen Chen (2005), "intensive and diversified emergy sources build up the structure and enhance metabolism in urban areas." (p. 49).

Mathematical Definition

To understand the concept of emergy it is first necessary to understand Exergy: the real proportion of the energy that can drive mechanical work.

E_x = (Gibbs free energy) + (gravitational potential energy) + (kinetic energy)

The Gibbs free energy is the available thermodynamic/chemical energy. Forms of energy such as radiation and thermal energy can not be converted completely to work, and have exergy content less than their energy content.Energy Memory, Energy in a Body

Energy Memory, Energy in a Body

Scienceman's inclusion of the term energy memory in the definition of the word emergy implies that the properties of physical-biological-chemical materials can be included within the domain of the emergy schema. In pronouncing the new nomenclature, Scienceman (1997, p. 210-211) wrote:

> "I now describe 'emergy' as meaning 'energy memory', meaning a measure of the quantity of original form energy [fJ] which has been totally used up or transformed into a new form of energy. The original form has disappeared and has become only a memory, a memory stored up in emergent properties and transformity."

Scienceman and El-Youssef (1993, p. 219), constructed the system of emergy units where the symbol for energy memory is a prefix, 'em-', which is expressed in terms of SI base units of p, meaning "past". In contrast, Bastianoni et al (2007) define energy memory as: "the embodiment or enfolding in processes or products of energy from different sources", with the same dimensions as energy [ML^2T-2] .

As it is used by Scienceman, the term "energy memory" implies a "memory algebra" (1997, p.211), which does not appear to obey the "energy conservation algebra" (Ibid.). This non-conservative view of systems is motivated by similar questions posed by Dirac. The energy conservation algebra simply states that the sum of the mass and energy must be conserved in any energy transformation, so that the sum of mass and energy in the universe remains constant forever (see for example Ohta 1994, p.3). While it is useful to assume that the sum of mass and energy in the universe remain constant forever, distributions of mass and energy in the universe do not remain constant forever. For example, the earth accumulates energy through the energy transformations carried out during photosynthesis. Another example is the hypertrophic growth of a heterotrophic organism. Consider also a simple closed electronic circuit with battery, and resistor. Because this circuit is closed, the quantity of electrons is constant. This is to say that the electrical form energy [e-fJ] is conserved. However the chemical form of energy that provides the force to move electrons in this circuit is not conserved. The heat energy form dissipates across the resistor as the chemical form of energy performs the work of moving electrons. Depending how one understands the concepts of open and closed systems, the circuit can be called closed to electron flow, but open to heat flow. In thermodynamics it is considered a closed system, but one that is isolated to electron flow, and not isolated to heat flow.

The memory algebra was designed to give a quantitative account of non-conservative energy transformations. The memory algebra also seems to have a connection with the physiological application of the cybernetic concept of "error" which is sometimes referred to as "memory". In addition to this H.T.Odum offered the theory that the toxic action of any substance is proportional to the emergy. This theory was partly based on research which found that pollution stress was greater on organisms with higher transformity (H.T.Odum 1994, p. 529). Further to this, G.P.Genoni, E.I.Meyer and A. Ulrich (2003) used the transformity measure to examine the hypothesis of a positive relationship between the "rarity" of a chemical element and its tendency to bioaccumulate, as energy in the body/entity under examination.

Empower and Maximum Empower

Empower refers to the flow rate of emergy: "The time rate of change of emergy is empower, analogous to the time rate of change of energy, power." (Scienceman 1987, p. 262.) Maximum empower therefore refers to the maximum flow rate of emergy. Considered as a principle, maximum empower has been proposed as a corollary of the maximum power principle, and is assumed to describe the organisational law of evolution. Accordingly H.T.Odum suggested that Lotka's maximum power principle be restated as the "Maximum Empower Principle"

The "Maximum Em-Power Principle" (Lotka-Odum) is generally considered as the "Fourth Thermodynamic Principle" (mainly) because of its practical validity for a very wide class of physical and biological systems (C.Giannantoni 2000, § 13, p. 155)

Citing H.T.Odum, J.L.Hau and B.R.Bakshi say that, "this principle determines which systems, ecological and economic, will survive over time and hence contribute to future systems" (2004, p.15).

Definition of the Maximum Empower Principle in Words

In the self-organizational process, systems develop those parts, processes, and relationships that maximize useful empower. (H.T. & E.C. Odum 2000, p. 71).

"The maximum empower principle is a unifying concept that explains why there are material cycles, autocatalytic feedbacks, successional stages, spatial concentrations in centers, and pulsing over time. Designs prevail that maximize empower". This is to say that "those elements or individuals whose patterns of action do not result in maximum production tend to be replaced eventually" (H.T. & E.C.Odum 2000, p.71). M.T.Brown and S.Ulgiati both maintain that "The total available emergy flow drives the system behaviour according to the Maximum Empower Principle, determining the size of the system itself and its growth rate." (2001, p. 109).

Mathematical Definition of the Maximum Empower Principle

As a corollary of maximum power efficiency, a mathematical statement of the maximum empower efficiency principle still needs to be clarified - see discussion

Emergy Acounting and Emergy "Analysis" Vs Emergy Synthesis

Emergy accounting is a GLOBAL method of accounting concerned with the input of solar energy equity at the global level. The emergy accounting methodology seeks to account for the energy used in developing energy of higher quality, which are capable of controlling ecosystem and economic functions. M.T.Brown and S.Ulgiati say that emergy accounting is a "method of valuation" that, "uses the thermodynamic basis of all forms of energy and materials, but converts them into equivalents of one form of energy, usually sunlight." (1999, p. 4). Emergy "analysis", therefore aims to identify how this energy, and its comparable inputs as fossil fuels etc., are distributed. Emergy accounting assumes three things:

1. That every sector of the world economy is, in the final analysis, dependent on the total global energy budget.
2. None of the sectors of the world economy overlap in their function.
3. Non-human sectors (i.e. non-human ecological systems) are included in the world economy

The sum result of these assumptions is that all sectors considered together produce a "complete" economy. It therefore includes energy, emergy, commodity and money flows. The method rests heavily on H.T.Odum's use of the "maximum power principle". In an evolutionary sense the use of this principle to analyse the world economy implies that the global economy will run close to optimal efficiency if and only if competition is unimpeded by cultural, communication, geographical, legal or other matters. However because emergy involves the combination of heterogeneous energy forms, D.M.Scienceman now only refers to Emergy Synthesis, preferring to see the notion of "emergy analysis" as an oxymoron.

Controversies

Acceptance of Nomenclature and Units

Corrado Giannantoni maintains that the maximum empower principle is generally accepted as the fourth thermodynamic principle (see above quote and reference). However the exchange between Scienceman (1997), H.T.Odum (1997), and Brown (1997), highlights the fact that various authors who have used the nomenclature are not agreement about how it is to be used - there appears to be significant differences of opinion regarding the physical, and scientific basis of the concept. To demonstrate this point, some authors such as Bastianoni et al. (2007, p.1159) writing that transformity is a dimensionless function, but others such as Odum (1994 p.251, 1988, p.1135) stating that it is not a dimensionless ratio, with Scienceman (1997, p.212) writing that:

"no one, to my knowledge has displayed any new dimensions of emergy or transformity in relation to the SI units, except my own provisional paper (Scienceman, 1993) which inserted a 'p' (meaning 'past') in the dimensions of relevant qualities. "

For some scientists, there are two fundamental requirements for something to be considered a thermodynamic or energetic principle. Firstly is an experimental technique, or instrument, used to quantitatively measure the phenomenon of interest. In the case of maximum empower this would mean the specification of an instrument that measures 'empower'. A second requirement is a set of mathematical equations that demonstrate, in the current case, an experimentally testable relationship of the phenomenon of "empower" to other thermodynamic variables.

A consequence of the first criteria is that serious scientific scholarship using the emergy nomenclature will need something like an "empower meter". Giannantoni has attempted to give the mathematics, however does not appear to have specified an empower meter by which to quantitatively measure empower. Although the concept of maximum empower has been used in peer-reviewed journals to model and quantify the ecological-economic sustainability of a region and nation, the question remains as to whether it qualifies as the 4th thermodynamic law, and apparently will not be resolved until an empower meter or equivalent is constructed. Given it is proposed as the 4th thermodynamic law acceptance of results would also probably only come more widely with general consensus in well respected physics journals like the The International Journal of Thermodynamics, or perhaps Physical Review. As long as physicists do not recognise it as the 4th thermodynamic law, it is unlikely scientists more generally will accept it and use the concept to unify disciplines like physics, biology and chemistry (not to mention society, economics and religion).

Value Theory

A controversial application of the concept is with respect to Value theory. For H.T. Odum, "embodied energy is a measure of value, in one of the meanings of the word 'value'" (H.T. Odum 1994, p.251–In this quote the term "embodied energy" is synonymous with "energy", see Embodied energy). H.T. Odum (1996) understood energy to encompass not only the above considerations, but also the Lumar notion of utility as a "donor-type value." As Scienceman (1987, p. 260) noted:

> "Odum ... clarified his theory of value: "Embodied energy was defined as a way to measure cumulative action of energies in chains and webs. Embodied energy provides an alternative theory of value, is useful for tracing sources, especially net energy, determining relative importance of components, and covering free items that are not covered by money." Of critical importance, Odum ... wrote: "An energy theory of value is based on embodied energy.... The energy transformation ratio, by giving the embodied energy per unit of actual energy, provides an intensive factor for value in the way that temperature is an intensive factor for heat".

This conception has not yet received wide support or critical analysis by value theorists. In fact it is this application which seems to act as a block to the wider appeal of Energy Synthesis in the sciences and arts. This could be due to the attitude which regards the arts, qualitative aesthetics, and spiritual systems of value as incommensurable with rational scientific systems of quantitative value measurement and prediction.

Environmental Ethics

Further controversial implications of this conception of value are in the fields of jurisprudence and policy where H.T. Odum suggests a consequentialist ethic. The kind of consequentialism employed is utilitarianism which seeks the quantitative maximization of value. Policies and laws that aim at long term sustainability are, according to this ethic, those that maximize value. From the perspective of energy synthesis, "value" is defined as "empower". Therefore, from the perspective of energy evaluation, policies and laws that aim at long term sustainability by maximizing value, consequently aim to maximize empower. Semantically speaking then, energy theorists imbue the concept of empower with both scientific, and normative content. In this context the division between social and physical science melts into air.

Emergy and Policy

In terms of energy, and environmental resource policy, H.T.Odum (1995b) maintained that, whether by trial and error or by intent, the maximum empower principle explains what is successful in self-organization. Odum and Odum (2001, p.71) claimed that the maximum empower principle accounts for many features of system performance and design:

> "It is a guideline for selecting policy. For policy the principle is: Choose alternatives that maximize empower intake and use."

Religion

As if matters were not sufficiently controversial, Scienceman (1995, p.1) added further fuel to the embers of confusion with the statement that :

"The concept of GOD is merely a 'personification' of emergy "

ALTERNATIVE ENERGY

Renewable energy sources, such as wind, flowing water, solar energy and biomass, which create less environmental damage and pollution than fossil fuels, and offer an alternative to nonrenewable resources.

The main article for this category is Alternative energy.

Alternative Energy is the by far smartest thing i've ever heard of. it's very tight as teens would say. ~Ethel Bennett~

Subcategories

There are 4 subcategories in this category, which are shown below. More may be shown on subsequent pages.

- Emergy
- Inertial confinement fusion research lasers
- Renewable energy
- Solar energy

There are 72 pages in this section of this category.

- Alternative energy
- Bernard S. Baker

- Biobased Product
- Bioconversion
- BlueEnergy
- Bubble fusion
- The Clean Tech Revolution
- Clostridium acetobutylicum
- ComES
- Comparison of automobile fuel technologies
- DEMO
- Deep ocean water
- Directive on the energy performance of buildings
- Dye-sensitized solar cells
- Electranet
- Energy Innovations
- Energy harvesting
- Envision Solar
- Fuel cell
- Fuel cell bus trial
- Future energy development
- Gamesa Corporación Tecnológica
- Geothermal power
- Gharo Wind Power Plant
- Ian Gilmartin
- Glanford Power Station

- GreenFuel Technologies Corporation
- Grid-tied electrical system
- Hydrogen energy plant in Denmark
- Industrial biotechnology
- Inertial fusion power plant
- Infinite Energy (magazine)
- Iowa Stored Energy Park
- ITER
- KSTAR
- Kilowatt Ours
- Laddermill
- Lovecraft Biofuels
- Eugene Mallove
- Methane clathrate
- Michigan Alternative and Renewable Energy Center
- Minewater Project
- New energy
- Ocean thermal energy conversion
- PACER
- Pacesetters
- Passive cooling
- Passive house
- Pelamis wave energy converter

- Plug Power
- Pyrolysis
- Renewable Portfolio Standard
- Renewable energy
- Selective surface
- SkySails
- Sustainable energy
- Tokamak
- Trisops
- User:Anthony717/sandbox
- Vegetable oil economy
- Vehicle-to-grid
- Wave farm
- Wave power
- Wind Over Water
- Wind farm
- Wind power
- Wind power in Asia
- Wind power in Ireland
- Wind-Diesel Hybrid Power Systems
- World Alliance for Decentralized Energy
- World energy resources and consumption
- Zero-energy building

RENEWABLE ENERGY

Renewable energy effectively utilizes natural resources such as sunlight, wind, tides and geothermal heat, which are naturally replenished.

Renewable energy technologies range from solar power, wind power, and hydroelectricity to biomass and biofuels for transportation. About 13 percent of primary energy comes from renewables, with most of this coming from traditional biomass like wood-burning.

Hydropower is the next largest source, providing 2-3%, and modern technologies like geothermal, wind, solar, and marine energy together produce less than 1% of total world energy demand.

The technical potential for their use is very large, exceeding all other readily available sources.

Renewable energy technologies are sometimes criticised for being unreliable or unsightly, yet the market is growing for many forms of renewable energy. Wind power has a worldwide installed capacity of 74,223 MW and is widely used in several European countries and the USA.

The manufacturing output of the photovoltaics industry reached more than 2,000 MW per year in 2006,and PV power plants are particularly popular in Germany. Solar thermal power stations operate in the USA and Spain, and the largest of these is the 354 MW SEGS power plant in the Mojave Desert.

The world's largest geothermal power installation is The Geysers in California, with a rated capacity of 750 MW. Brazil has one of the largest renewable energy programs in the world, involving production of ethanol fuel from sugar cane, and ethanol now provides 18 percent of the country's automotive fuel. Ethanol fuel is also widely available in the USA.

While there are many large-scale renewable energy projects, renewable technologies are also suited to small off-grid applications, sometimes in rural and remote areas, where energy is often crucial in human development.

Kenya has the world's highest household solar ownership rate with roughly 30,000 small (20-100 watt) solar power systems sold per year.

Climate change concerns coupled with high oil prices, peak oil and increasing government support are driving increasing renewable energy legislation, incentives and commercialization. EU leaders reached agreement in principle in March that 20 percent of the bloc's energy should be produced from renewable fuels by 2020, as part of its drive to cut emissions of carbon dioxide, blamed in part for global warming. Investment capital flowing into renewable energy climbed from $80 billion in 2005 to a record $100 billion in 2006. Some very large corporations such as BP, GE, Sharp, and Shell are investing in the renewable energy sector.

The U.S. renewable energy and energy efficiency industries employed 8 million people in 2006, while generating 933 billion dollars in sales, $100 billion in profits, and $150 billion in increased federal, state, and local government tax revenues, according to a new report from the American Solar Energy Society (ASES). The report notes that it's difficult to define the energy efficiency industry, but even focusing on the renewable energy industry, it found 196,000 people directly employed by the industry, a total of 452,000 jobs created, and revenues of $39.2 billion in 2006.

Scenarios

Looking forward to 2030, the American Solar Energy Society report examines three scenarios: a "business as usual" scenario, with no major policy changes; a moderate scenario that includes incremental policy advances; and an advanced scenario of aggressive growth in renewable energy and energy efficiency.

In the "business as usual" scenario, the jobs created by renewable energy increase 190% by 2030, while jobs created by energy efficiency increase by 85%.

In the moderate scenario, the jobs created by renewable energy increase nearly 7-fold, while jobs created by energy efficiency more than double.

And in the advanced scenario, the jobs created by renewable energy increase 17-fold, while jobs created by energy efficiency quadruple. In the advanced scenario, renewable energy revenues increase to nearly $600 billion, while energy efficiency revenues increase to almost $4 trillion.

Main Renewable Energy Technologies

The majority of renewable energy technologies are directly or indirectly powered by the sun. The Earth-Atmosphere system is in equilibrium such that heat radiation into space is equal to incoming solar radiation, the resulting level of energy within the Earth-Atmosphere system can roughly be described as the Earth's "climate." The hydrosphere (water) absorbs a major fraction of the incoming radiation. Most radiation is absorbed at low latitudes around the equator, but this energy is dissipated around the globe in the form of winds and ocean currents. Wave motion may play a role in the process of transferring mechanical energy between the atmosphere and the ocean through wind stress. Solar energy is also responsible for the distribution of precipitation which is tapped by hydroelectric projects, and for the growth of plants used to create biofuels.

Renewable energy flows involve natural phenomena such as sunlight, wind, tides and geothermal heat, as the International Energy Agency explains:

"Renewable energy is derived from natural processes that are replenished constantly. In its various forms, it derives directly from the sun, or from heat generated deep within the earth. Included in the definition is electricity and heat generated from solar, wind, ocean, hydropower, biomass, geothermal resources, and biofuels and hydrogen derived from renewable resources."

Each of these sources has unique characteristics which influence how and where they are used.

Wind Power

Airflows can be used to run wind turbines. Modern wind turbines range from around 600kW to up to 5 MW of rated power, although turbines, with rated output of 1.5-3 MW, have become the most common for commercial use; the power output of a turbine is a function of the cube of the wind speed, so as wind speed increases, power output increases dramatically. Areas where winds are stronger and more constant, such as offshore and high altitude sites, are preferred locations for wind farms.

Wind power is the fastest growing of the renewable energy technologies, though it currently provides less than 0.5% of global energy. Over the past decade, global installed maximum capacity increased from 2,500 MW in 1992 to just over 40,000 MW at the end of 2003, at an annual growth rate of near 30%. As wind power has become more prominent and viable, several public schools are incorporating sustainable wind power into the energy grid of their school in order to cut power costs. Due to the intermittency of wind resources, most deployed turbines in the EU produce electricity an average of 25% of the hours in a year (a capacity factor of 25%), but under favourable wind regimes some reach 35% or higher. Capacity factors are a function of seasonal wind fluctuations and may be higher in winter. It would mean that a typical 5 MW turbine in the EU would have an average output of 1.7 MW.

Globally, the long-term technical potential of wind energy is believed to be five times total current global energy production, or 40 times current electricity demand. This could require large amounts of land to be utilized for wind turbines, particularly in areas of higher wind resources. Offshore resources experience mean wind speeds of ~90% greater than that of land, so offshore resources could contribute substantially more energy. This number could also increase with higher altitude ground-based or airborne wind turbines

Wind strengths near the Earth's surface vary and thus cannot guarantee continuous power unless combined with other energy sources or storage systems. Some estimates suggest that 1,000 MW of conventional wind generation capacity can be relied on for just 333 MW of continuous power. While this might change as technology evolves, advocates have suggested incorporating wind power with other power sources, or the use of energy storage techniques, with this in mind. It is best used in the context of a system that has significant reserve capacity such as hydro, or reserve load, such as a desalination plant, to mitigate the economic effects of resource variability.

Wind power is renewable and produces no greenhouse gases during operation, such as carbon dioxide and methane.

Water Power

Energy in water (in the form of motive energy or temperature differences) can be harnessed and used. Since water is about 800 times denser than air, even a slow flowing stream of water, or moderate sea swell, can yield considerable amounts of energy.

There are many forms of water energy:

- Hydroelectric energy is a term usually reserved for large-scale hydroelectric dams.
- Micro hydro systems are hydroelectric power installations that typically produce up to 100 kW of power. They are often used in water rich areas as a Remote Area Power Supply (RAPS). There are many of these installations around the world, including several delivering around 50 kW in the Solomon Islands.
- Wave power uses the energy in waves. The waves will usually make large pontoons go up and down in the water, leaving an area with reduced wave height in the "shadow". Wave power has now reached commercialization.
- Tidal power captures energy from the tides in a vertical direction. Tides come in, raise water levels in a basin, and tides roll out. Around low tide, the water in the basin is discharged through a turbine.
- In November 2007, British company Lunar Energy announced that, in conjunction with E.On, they would be building the world's first tidal energy farm off the coast of Pembrokshire in Wales. it will be the world's first deep-sea tidal-energy farm and will provide electricity for 5,000 homes. Eight underwater turbines, each 25 metres long and 15 metres high, are to be installed on the sea bottom off St David's peninsula. Construction is due to start in the summer of 2008 and the proposed tidal energy turbines, described as "a wind farm under the sea", should be operational by 2010.

- Tidal stream power captures energy from the flow of tides, usually using underwater plant resembling a small wind turbine. Tidal stream power demonstration projects exist, and the first commercial prototype will be installed in Strangford Lough in September 2007.
- Ocean thermal energy conversion (OTEC) uses the temperature difference between the warmer surface of the ocean and the colder lower recesses. To this end, it employs a cyclic heat engine. OTEC has not been field-tested on a large scale.
- Deep lake water cooling, although not technically an energy generation method, can save a lot of energy in summer. It uses submerged pipes as a heat sink for climate control systems. Lake-bottom water is a year-round local constant of about 4 °C.
- Blue energy is the reverse of desalination. This form of energy is in research.

Solar Energy Use

In this context, "solar energy" refers to energy that is collected from sunlight. Solar energy can be applied in many ways, including to:

- Generate electricity using photovoltaic solar cells.
- Generate electricity using concentrated solar power.
- Generate electricity by heating trapped air which rotates turbines in a Solar updraft tower.
- Heat buildings, directly, through passive solar design.
- Heat foodstuffs, through solar ovens.
- Heat water or air for domestic hot water and space heating needs using solar-thermal panels.
- Heat and cool air through use of solar chimneys.
- Generate electricity in geosynchronous orbit using solar power satellites.

Biofuel

Plants use photosynthesis to grow and produce biomass. Also known as biomatter, biomass can be used directly as fuel or to produce liquid biofuel. Agriculturally produced biomass fuels, such as biodiesel, ethanol and bagasse (often a by-product of sugar cane cultivation) can be burned in internal combustion engines or boilers. Typically biofuel is burned to release its stored chemical energy. Research into more efficient methods of converting biofuels and other fuels into electricity utilizing fuel cells is an area of very active work.

Liquid Biofuel

Liquid biofuel is usually either a bioalcohol such as ethanol or a bio-oil such as biodiesel and straight vegetable oil. Biodiesel can be used in modern diesel vehicles with little or no modification to the engine and can be made from waste and virgin vegetable and animal oil and fats (lipids). Virgin vegetable oils can be used in modified diesel engines. In fact the Diesel engine was originally designed to run on vegetable oil rather than fossil fuel. A major benefit of biodiesel is lower emissions. The use of biodiesel reduces emission of carbon monoxide and other hydrocarbons by 20 to 40%. In some areas corn, cornstalks, sugarbeets, sugar cane, and switchgrasses are grown specifically to produce ethanol (also known as grain alcohol) a liquid which can be used in internal combustion engines and fuel cells. Ethanol is being phased into the current energy infrastructure. E85 is a fuel composed of 85% ethanol and 15% gasoline that is sold to consumers. Biobutanol is being developed as an alternative to bioethanol.

In the future, there might be bio-synthetic liquid fuel available. It can be produced by the Fischer-Tropsch process, also called Biomass-To-Liquids (BTL).

Solid Biomass

Direct use is usually in the form of combustible solids, either wood, the biogenic portion of municipal solid waste or combustible field crops. Field crops may be grown specifically for combustion or may be used for other purposes, and the processed plant waste then used for combustion. Most sorts of biomatter, including dried manure, can actually be burnt to heat water and to drive turbines.

Sugar cane residue, wheat chaff, corn cobs and other plant matter can be, and are, burned quite successfully. The net carbon dioxide emissions that are added to the atmosphere by this process are only from the fossil fuel that is often currently consumed to plant, fertilize, harvest and transport the biomass.

Processes to harvest biomass from short-rotation poplars and willows, and perennial grasses such as switchgrass, phalaris, and miscanthus, require less frequent cultivation and less nitrogen than from typical annual crops. Pelletizing miscanthus and co-firing it with coal for generating electricity is being studied and may be economically viable. The higher heating value of cellulose is about 17.4 MJ/kg . The estimated yield of ethanol from dry cellulose is about 0.2 kg of ethanol per kg of cellulose (60 gal/ton). Since the higher heating value of ethanol is 29.7 MJ/kg of ethanol it would be 5.94 MJ/kg of the cellulose that it is made from. Thus the ethanol contains only about 1/3 as much energy as the cellulose that it was made from. Co-firing cellulose with coal would replace about three times as much fossil fuel as using the cellulose to make ethanol. The replaced coal would produce 0.0946 kg CO_2 /MJ while the replaced liquid fuel would produce only about 0.0733 kg CO_2 /MJ so co-firing the cellulose with coal is about 3.8 times more effective at reducing CO_2 emissions than using it to make ethanol.

Solid biomass can also be gasified, and used as described in the next section.

Biogas

Biogas can easily be produced from current waste streams, such as: paper production, sugar production, sewage, animal waste and so forth. These various waste streams have to be slurried together and allowed to naturally ferment, producing methane gas. This can be done by converting current sewage plants into biogas plants. When a biogas plant has extracted all the methane it can, the remains are sometimes better suitable as fertilizer than the original biomass.

Alternatively biogas can be produced via advanced waste processing systems such as mechanical biological treatment. These systems recover the recyclable elements of household waste and process the biodegradable fraction in anaerobic digesters.

Renewable natural gas is a biogas which has been upgraded to a quality similar to natural gas. By upgrading the quality to that of natural gas, it becomes possible to distribute the gas to the mass market via gas grid.

Geothermal Energy

Geothermal energy is energy obtained by tapping the heat of the earth itself, usually from kilometers deep into the Earth's crust. It is expensive to build a power station but operating costs are low resulting in low energy costs for suitable sites. Ultimately, this energy derives from heat in the Earth's core. The government of Iceland states: "It should be stressed that the geothermal resource is not strictly renewable in the same sense as the hydro resource." It estimates that Iceland's geothermal energy could provide 1700 MW for over 100 years, compared to the current production of 140 MW. The International Energy Agency classifies geothermal power as renewable.

Three types of power plants are used to generate power from geothermal energy: dry steam, flash, and binary. Dry steam plants take steam out of fractures in the ground and use it to directly drive a turbine that spins a generator. Flash plants take hot water, usually at temperatures over 200 °C, out of the ground, and allows it to boil as it rises to the surface then separates the steam phase in steam/water separators and then runs the steam through a turbine. In binary plants, the hot water flows through heat exchangers, boiling an organic fluid that spins the turbine. The condensed steam and remaining geothermal fluid from all three types of plants are injected back into the hot rock to pick up more heat.

The geothermal energy from the core of the Earth is closer to the surface in some areas than in others. Where hot underground steam or water can be tapped and brought to the surface it may be used to generate electricity. Such geothermal power sources exist in certain geologically unstable parts of the world such as Iceland, New Zealand, United States, the Philippines and Italy. The two most prominent areas for this in the United States are in the Yellowstone basin and in northern California. Iceland produced 170 MW geothermal power and heated 86% of all houses in the year 2000 through geothermal energy. Some 8000 MW of capacity is operational in total.

There is also the potential to generate geothermal energy from hot dry rocks. Holes at least 3 km deep are drilled into the earth. Some of these holes pump water into the earth, while other holes pump hot water out. The heat resource consists of hot underground radiogenic granite rocks, which heat up when there is enough sediment between the rock and the earths surface. Several companies in Australia are exploring this technology.

Renewable Energy Commercialization

Costs

Renewable energy systems encompass a broad, diverse array of technologies, and the current status of these can vary considerably. Some technologies are already mature and economically competitive (e.g. geothermal and hydropower), others need additional development to become competitive without subsidies. This can be helped by improvements to sub-components, such as electric generators.

The table shows an overview of costs of various renewable energy technologies. For comparison with the prices in the table, electricity production from a conventional coal-fired plant costs about 4¢/kWh. Though in some G8 nations the cost can be significantly higher at 7.88p (~15¢/kWh). Achieving further cost reductions as indicated in the table below requires further technology development, market deployment, and an increase in production capacities to mass production levels.

	2001 energy costs	*Potential future energy cost*
	Electricity	
Wind	4-8 ¢/kWh	3-10 ¢/kWh
Solar photovoltaic	25-160 ¢/kWh	5-25 ¢/kWh
Solar thermal	12-34 ¢/kWh	4-20 ¢/kWh
Large hydropower	2-10 ¢/kWh	2-10 ¢/kWh

Small hydropower	2-12 ¢/kWh	2-10 ¢/kWh
Geothermal	2-10 ¢/kWh	1-8 ¢/kWh
Biomass	3-12 ¢/kWh	4-10 ¢/kWh
Coal (comparison)	4¢/kWh	
	Heat	
Geothermal heat	0.5-5 ¢/kWh	0.5-5 ¢/kWh
Biomass - heat	1-6 ¢/kWh	1-5 ¢/kWh
Low temp solar heat	2-25 ¢/kWh	2-10 ¢/kWh
All costs are in 2001 $-cent per kilowatt-hour.		

Source: World Energy Assessment, 2004 update[33]

Wind Power Market Increase

Figures from the Global Wind Energy Council (GWEC) show that 2006 recorded an increase in installed wind power capacity of 15,197 megawatts (MW), taking the total installed capacity to 74,223 MW, up from 59,091 MW in 2005. Despite constraints facing supply chains for wind turbines, the annual market for wind continued to increase at the rate of 32% following the 2005 record year, in which the market grew by 41%. In terms of economic value, the wind energy sector has become one of the important players in the energy markets, with the total value of new generating equipment installed in 2006 reaching •18 billion, or US$23 billion.

The countries with the highest total installed capacity are Germany (20,621 MW), Spain (11,615 MW), the USA (11,603 MW), India (6,270 MW) and Denmark (3,136 MW). In terms of new installed capacity in 2006, the USA lead with 2,454 MW, followed by Germany (2,233 MW), India (1,840 MW), Spain (1,587 MW), China (1,347 MW) and France (810 MW).

In the UK, a licence to build the world's largest offshore windfarm, in the Thames estuary, has been granted. The London Array windfarm, 12 miles off Kent and Essex, should eventually consist of 341 turbines, occupying an area of 90 square miles. This is a £1.5 billion, 1,000 megawatt project, which will power one-third of London homes. The windfarm will produce an amount of energy that, if generated by conventional means, would result in 1.9 million tonnes of carbon dioxide emissions every year. It could also make up to 10% of the Government's 2010 renewables target.

New Generation of Solar Thermal Plants

Construction of the largest solar thermal power plant to be built in 15 years, in Boulder City, Nevada, is nearly complete. The 64MW Nevada Solar One power plant will generate enough power to meet the electricity needs of about 40,000 households and follows in the steps of the 354MW SEGS solar thermal power plants located in California's Mojave Desert. While California's solar plants have generated billions of kilowatt hours of electricity for the past two decades, the Nevada Solar One plant will use new technologies to capture even more energy from the sun.

The California Solar Initiative

As part of Governor Arnold Schwarzenegger's Million Solar Roofs Program, California has set a goal to create 3,000 megawatts of new, solar-produced electricity by 2017 - moving the state toward a cleaner energy future and helping lower the cost of solar systems for consumers. This is a comprehensive $2.8 billion program.

The California Solar Initiative offers cash incentives on solar PV systems of up to $2.50 a watt. These incentives, combined with federal tax incentives, can cover up to 50% of the total cost of a solar panel system.There are many financial incentives to support the use of renewable energy in other US states.

World's Largest Photovoltaic Power Plants

Construction of a 40 MW solar generation power plant is underway in the Saxon region of Germany. The Waldpolenz Solar Park will consist of some 550,000 thin-film solar modules. The direct current produced in the modules will be converted into alternating current and fed completely into the power grid. Once completed in 2009, the project will be one of the largest photovoltaic projects ever constructed. Currently the biggest PV plant in the world has an output capacity of around 12 megawatts.

A large photovoltaic power project has been completed in Portugal. The Serpa solar power plant is located at one of Europe's sunniest areas.The 11 megawatt plant covers 150 acres and is comprised of 52,000 PV panels. The panels are raised 2 metres off the ground and the area will remain productive grazing land. The project will provide enough energy for 8,000 homes and will save an estimated 30,000 tonnes of carbon dioxide emissions per year.

A $420 million large-scale Solar power station in Victoria is to be the biggest and most efficient solar photovoltaic power station in the world. Australian company Solar Systems will demonstrate its unique design incorporating space technology in a 154MW solar power station connected to the national grid. The power station will have the capability to concentrate the sun by 500 times onto the solar cells for ultra high power output. The Victorian power station will generate clean electricity directly from the sun to meet the annual needs of over 45,000 homes with zero greenhouse gas emissions.

However, when it comes to renewable energy systems and PV, it is not just large systems that matter. Building-integrated photovoltaics or "onsite" PV systems have the advantage of being matched to end use energy needs in terms of scale. So the energy is supplied close to where it is needed.

Use of Ethanol for Transportation

Brazil has one of the largest renewable energy programs in the world, involving production of ethanol fuel from sugar cane, and ethanol now provides 18 percent of the country's automotive fuel. As a partial result, Brazil, which years ago had to import a large share of the petroleum needed for domestic consumption, recently reached complete self-sufficiency in oil.

Most cars on the road today in the U.S. can run on blends of up to 10% ethanol, and motor vehicle manufacturers already produce vehicles designed to run on much higher ethanol blends. Ford, DaimlerChrysler, and GM are among the automobile companies that sell "flexible-fuel" cars, trucks, and minivans that can use gasoline and ethanol blends ranging from pure gasoline up to 85% ethanol (E85). By mid-2006, there were approximately six million E85-compatible vehicles on U.S. roads. The challenge is to expand the market for biofuels beyond the farm states where they have been most popular to date. Flex-fuel vehicles are assisting in this transition because they allow drivers to choose different fuels based on price and availability. The *Energy Policy Act of 2005*, which calls for 7.5 billion gallons of biofuels to be used annually by 2012, will also help to expand the market.

Wave Farms Expansion

Portugal now has the world's first commercial wave farm, the *Aguçadora Wave Park*, established in 2006. The farm will initially use three Pelamis P-750 machines generating 2.25 MW. Initial costs are put at •8.5 million. Subject to successful operation, a further •70 million is likely to be invested before 2009 on a further 28 machines to generate 525 MW.

Funding for a wave farm in Scotland was announced in February, 2007 by the Scottish Executive, at a cost of over 4 million pounds, as part of a £13 million funding packages for ocean power in Scotland. The farm will be the world's largest with a capacity of 3MW generated by four Pelamis machines.

Geothermal Energy Prospects

By the end of 2005 worldwide use of geothermal energy for electricity had reached 9.3 GWs, with an additional 28 GW used directly for heating. If heat recovered by ground source heat pumps is included, the non-electric use of geothermal energy is estimated at more than 100 GWt (gigawatts of thermal power) and is used commercially in over 70 countries.(sec 1.2) During 2005 contracts were placed for an additional 0.5 GW of capacity in the United States, while there were also plants under construction in 11 other countries.

Future Potential

While at present renewable energy sources only supply a fraction of current energy use (ca. 14% of primary energy use, mostly from traditional biomass), there is much potential that could be exploited in the future. As the table below illustrates, the technical potential of renewable energy sources is more than 18 times current global primary energy use and furthermore several times higher than projected energy use in 2100.

The Renewable Energy Resource Base (Exajoules a Year)

	Current use (2001)	*Technical potential*	*Theoretical potential*
Hydropower	9	50	147
Biomass energy	50	>276	2,900
Solar energy	0.1	>1,575	3,900,000
Wind energy	0.12	640	6,000
Geothermal energy	0.6	5,000	140,000,000
Ocean energy	not estimated	not estimated	7,400
Total	60	>7,600	>144,000,000

Current use is in primary energy equivalent.For comparison, the current global primary energy use (2001) is 402 Exajoules a year.

Source: World Energy Assessment 2001.

There are many different ways to assess potentials. The theoretical potential indicates the amount of energy theoretically available for energy purposes, such as, in the case of solar energy, the amount of incoming radiation at the earth's surface. The technical potential is a more practical estimate of how much could be put to human use by considering conversion efficiencies of the available technology and available land area. To give an idea of the constraints, the estimate for solar energy assumes that 1% of the world's unused land surface is used for solar power.

The technical potentials generally do not include economic or other environmental constraints, and the potentials that could be realized at an economically competitive level under current conditions and in a short time-frame is lower still.

Trends Favoring Renewables

The renewable market will boom when cost efficiency attains parity with other competing energy sources. The following trends are a few examples by which the renewables market is being helped to attain critical mass so that it becomes competitive enough vs fossil fuels:

Other than market forces, renewable industry often needs government sponsorship to help generate enough momentum in the market. Many countries and states have implemented incentives - like government tax subsidies, partial copayment schemes and various rebates over purchase of renewables - to encourage consumers to shift to renewable energy sources. Government grants fund for research in renewable technology to make the production cheaper and generation more efficient.

Development of loan programs that stimulate renewable favoring market forces with attractive return rates, buffer intial deployment costs and entice consumers to consider and purchase renewable technology. A famous example is the solar loan program sponsored by UNEP helping 100000 people finance solar power systems in India. Success in India's solar program has led to similar projects in other parts of developing world like Tunisia, Morocco, Indonesia and Mexico.

Imposition of high fossil fuel consumption / carbon taxes, and channel the revenue earned towards renewable energy development.

Many think-tanks are warning that the world needs an urgency driven concerted effort to create a competitive renewable energy infrastructure and market. The developed world can make more research investments to find better cost efficient technologies, and manufacturing could be transferred to developing countries in order to use low labor costs. The renewable energy market could increase fast enough to replace and initiate the decline of fossil fuel dominance and the world could then avert the looming climate and peak oil crises.

Most importantly, renewables is gaining credence among private investors as having the potential to grow into the next big industry. Many companies and venture capitalists are investing in photovoltaic development and manufacturing. This trend is particularly visible in Silicon valley, California, Europe, Japan.

Constraints and Opportunities

Critics suggest that some renewable energy applications may create pollution, be dangerous, take up large amounts of land, or be incapable of generating a large net amount of energy. Proponents advocate the use of "appropriate renewables", also known as soft energy technologies, as these have many advantages.

Availability

There is no shortage of solar-derived energy on Earth. Indeed the storages and flows of energy on the planet are very large relative to human needs.

- The amount of solar energy intercepted by the Earth every minute is greater than the amount of energy the world uses in fossil fuels each year.
- Tropical oceans absorb 560 trillion gigajoules (GJ) of solar energy each year, equivalent to 1,600 times the world's annual energy use.
- The energy in the winds that blow across the United States each year could produce more than 16 billion GJ of electricity—more than one and one-half times the electricity consumed in the United States in 2000.
- Annual photosynthesis by the vegetation in the United States is 50 billion GJ, equivalent to nearly 60% of the nation's annual fossil fuel use.

A criticism of some renewable sources is their intermittent nature. But a variety of renewable sources in combination can overcome this problem. As Amory Lovins explains:

"Stormy weather, bad for direct solar collection, is generally good for windmills and small hydropower plants; dry, sunny weather, bad for hydropower, is ideal for photovoltaics."

The challenge of variable power supply may be further alleviated by energy storage. Available storage options include pumped-storage hydro systems, batteries, hydrogen fuel cells,, and thermal mass. Initial investments in such energy storage systems can be high, although the costs can be recovered over the life of the system.

Wave energy is continuously available, although wave intensity varies by season. A wave energy scheme installed in Australia generates electricity with an 80% availability factor.

Aesthetics

Both solar and wind generating stations have been criticized from an aesthetic point of view. However, methods and opportunities exist to deploy these renewable technologies efficiently and unobtrusively: fixed solar collectors can double as noise barriers along highways, and extensive roadway, parking lot, and roof-top area is currently available; amorphous photovoltaic cells can also be used to tint windows and produce energy. Advocates of renewable energy also argue that current infrastructure is less aethetically pleasing than alternatives, but sited further from the view of most critics.

Environmental and Social Considerations

While most renewable energy sources do not produce pollution directly, the materials, industrial processes, and construction equipment used to create them may generate waste and pollution. Some renewable energy systems actually create environmental problems. For instance, older wind turbines can be hazardous to flying birds.

Land Area Required

Another environmental issue, particularly with biomass and biofuels, is the large amount of land required to harvest energy, which otherwise could be used for other purposes or left as undeveloped land. However, it should be pointed out that these fuels may reduce the need for harvesting non-renewable energy sources, such as vast strip-mined areas and slag mountains for coal, safety zones around nuclear plants, and hundreds of square miles being strip-mined for oil sands. These responses, however, do not account for the extremely high biodiversity and endemism of land used for ethanol crops, particularly sugar cane.

In the U.S., crops grown for biofuels are the most land- and water-intensive of the renewable energy sources. In 2005, about 12% of the nation's corn crop (covering 11 million acres (45,000 km^2) of farmland) was used to produce four billion gallons of ethanol—which equates to about 2% of annual U.S. gasoline consumption. For biofuels to make a much larger contribution to the energy economy, the industry will have to accelerate the development of new feedstocks, agricultural practices, and technologies that are more land and water efficient. Already, the efficiency of biofuels production has increased significantly and there are new methods to boost biofuel production.

Hydroelectric Dams

The major advantage of hydroelectric systems is the elimination of the cost of fuel. Other advantages include longer life than fuel-fired generation, low operating costs, and the provision of facilities for water sports. Operation of pumped-storage plants improves the daily load factor of the generation system. Overall, hydroelectric power can be far less expensive than electricity generated from fossil fuels or nuclear energy, and areas with abundant hydroelectric power attract industry.

However, there are several major disadvantages of hydroelectric systems. These include: dislocation of people living where the reservoirs are planned, release of significant amounts of carbon dioxide at construction and flooding of the reservoir, disruption of aquatic ecosystems and birdlife, adverse impacts on the river environment, potential risks of sabotage and terrorism, and in rare cases catastrophic failure of the dam wall. (See Hydroelectricity article for details.)

Hydroelectric power is now more difficult to site in developed nations because most major sites within these nations are either already being exploited or may be unavailable for other reasons such as environmental considerations.

Wind Farms

A wind farm, when installed on agricultural land, has one of the lowest environmental impacts of all energy sources:

- It occupies less land area per kilowatt-hour (kWh) of electricity generated than any other energy conversion system, apart from rooftop solar energy, and is compatible with grazing and crops.
- It generates the energy used in its construction in just 3 months of operation, yet its operational lifetime is 20-25 years.
- Greenhouse gas emissions and air pollution produced by its construction are tiny and declining. There are no emissions or pollution produced by its operation.
- In substituting for base-load coal power, wind power produces a net decrease in greenhouse gas emissions and air pollution, and a net increase in biodiversity.
- Modern wind turbines are almost silent and rotate so slowly (in terms of revolutions per minute) that they are rarely a hazard to birds.

Studies of birds and offshore wind farms in Europe have found that there are very few bird collisions. Several offshore wind sites in Europe have been in areas heavily used by seabirds. Improvements in wind turbine design, including a much slower rate of rotation of the blades and a smooth tower base instead of perchable lattice towers, have helped reduce bird mortality at wind farms around the world. However older smaller wind turbines may be hazardous to flying birds. Birds are severely impacted by fossil fuel energy; examples include birds dying from exposure to oil spills, habitat loss from acid rain and mountaintop removal coal mining, and mercury poisoning.

Longevity Issues

Though a source of renewable energy may last for billions of years, renewable energy infrastructure, like hydroelectric dams, will not last forever, and must be removed and replaced at some point. Events like the shifting of riverbeds, or changing weather patterns could potentially alter or even halt the function of hydroelectric dams, lowering the amount of time they are available to generate electricity.

Although geothermal sites are capable of providing heat for many decades, eventually specific locations may cool down. It is likely that in these locations, the system was designed too large for the site, since there is only so much energy that can be stored and replenished in a given volume of earth. Some interpret this as meaning a specific geothermal location can undergo depletion.

Biofuels Production

All biomass needs to go through some of these steps: it needs to be grown, collected, dried, fermented and burned. All of these steps require resources and an infrastructure.

Some studies contend that ethanol is "energy negative", meaning that it takes more energy to produce than is contained in the final product. However, a large number of recent studies, including a 2006 article in the journal *Science* offer the opinion that fuels like ethanol are energy positive. Furthermore, fossil fuels also require significant energy inputs which have seldom been accounted for in the past.

Additionally, ethanol is not the only product created during production, and the energy content of the by-products must also be considered. Corn is typically 66% starch and the remaining 33% is not fermented. This unfermented component is called distillers grain, which is high in fats and proteins, and makes good animal feed. In Brazil, where sugar cane is used, the yield is higher, and conversion to ethanol is somewhat more energy efficient than corn. Recent developments with cellulosic ethanol production may improve yields even further.

According to the International Energy Agency, new biofuels technologies being developed today, notably cellulosic ethanol, could allow biofuels to play a much bigger role in the future than previously thought. Cellulosic ethanol can be made from plant matter composed primarily of inedible cellulose fibers that form the stems and branches of most plants. Crop residues (such as corn stalks, wheat straw and rice straw), wood waste, and municipal solid waste are potential sources of cellulosic biomass. Dedicated energy crops, such as switchgrass, are also promising cellulose sources that can be sustainably produced in many regions of the United States.

The ethanol and biodiesel production industries also create jobs in plant construction, operations, and maintenance, mostly in rural communities. According to the Renewable Fuels Association, the ethanol industry created almost 154,000 U.S. jobs in 2005 alone, boosting household income by $5.7 billion. It also contributed about $3.5 billion in tax revenues at the local, state, and federal levels.

Diversification

The U.S. electric power industry now relies on large, central power stations, including coal, natural gas, nuclear, and hydropower plants that together generate more than 95% of the nation's electricity. Over the next few decades uses of renewable energy could help to diversify the nation's bulk power supply. Already, appropriate renewable resources (which excludes large hydropower) produce 12% of northern California's electricity.

Although most of today's electricity comes from large, central-station power plants, new technologies offer a range of options for generating electricity nearer to where it is needed, saving on the cost of transmitting and distributing power and improving the overall efficiency and reliability of the system.

Improving energy efficiency represents the most immediate and often the most cost-effective way to reduce oil dependence, improve energy security, and reduce the health and environmental impact of the energy system. By reducing the total energy requirements of the economy, improved energy efficiency could make increased reliance on renewable energy sources more practical and affordable.

Other Issues

Nuclear Power

In 1983, physicist Bernard Cohen proposed that uranium could provide effectively inexhaustible energy.

We thus conclude that all the world's energy requirements for the remaining 5×10^9 yr of existence of life on Earth could be provided by breeder reactors without the cost of electricity rising by as much as 1% due to fuel costs. This is consistent with the definition of a "renewable" energy source in the sense in which that term is generally used.

Nuclear fusion, if developed, could have a similarly large potential and is expected to have fewer waste and waste-containment issues than fission.

Neither nuclear fission nor nuclear fusion are generally regarded as forms of renewable energy.

Fossil Fuels

Fossil fuels consist of deposits of once living organisms that took centuries to form. The energy comes mostly from hydrogen and carbon bonds. The non-renewability of these sources will most likely cause prices to rise up to a point where they are no longer economically feasible. This is to say that the Hubbert peak theory applies to all fossil fuels.

Renewable Electricity

Renewable electricity is a term used for electricity primarily produced from the renewable resources (solar, wind... energy). It is generally labeled as green energy.

Transmission

If renewable and distributed generation were to become widespread, electric power transmission and electricity distribution systems might no longer be the main distributors of electrical energy but would operate to balance the electricity needs of local communities. Those with surplus energy would sell to areas needing "top ups". That is, network operation would require a shift from 'passive management' — where generators are hooked up and the system is operated to get electricity 'downstream' to the consumer — to 'active management', wherein generators are spread across a network and inputs and outputs need to be constantly monitored to ensure proper balancing occurs within the system. Some governments and regulators are moving to address this, though much remains to be done. One potential solution is the increased use of active management of electricity transmission and distribution networks. This will require significant changes in the way that such networks are operated.

However, on a smaller scale, use of renewable energy produced on site reduces burdens on electricity distribution systems. Current systems, while rarely economically efficient, have shown that an average household with an appropriately-sized solar panel array and energy storage system needs electricity from outside sources for only a few hours per week. By matching electricity supply to end-use needs, advocates of renewable energy and the soft energy path believe electricity systems will become smaller and easier to manage, rather than the opposite (see Soft energy technology).

Market Development of Renewable Heat Energy

Renewable heat is the generation of heat from renewable sources. Much current discussion on renewable energy focuses on the generation of electrical energy, despite the fact that many colder countries consume more energy for heating than as electricity. The United Kingdom consumes 350 TWh of electric power, and 840 TWh of gas and other fuels for heating annually. The residential sector alone consumes a massive 550 TWh of energy for heating, mainly in the form of gas.

Renewable electric power is becoming cheap and convenient enough to place it, in many cases, within reach of the average consumer. By contrast, the market for renewable heat is mostly inaccessible to domestic consumers due to inconvenience of supply, and high capital costs. Heating accounts for a large proportion of energy consumption, however a universally accessible market for renewable heat is yet to emerge. Solutions such as geothermal heat pumps may be more widely applicable, but may not be economical in all cases. Also see renewable energy development.

Renewable Energy in Pop Culture

The FLL(First Lego League)in the year 2007-2008 the theme is Alternative energy. the teams try to acomplish goals of switching normal cars with fuel cell cars, putting solar panels on houses etc. The teams must also do an energy audit of a building of their choice. They must write a play, skit, song or other type of media concerning the audit and how to make the building more energy efficient.

BIOFUEL

Biofuel (also called agrofuel) can be broadly defined as solid, liquid, or gas fuel consisting of, or derived from biomass. This article, however, is principally about biofuel in the form of liquid or gas transportation fuel derived from biomass. Biomass can also be used directly for heating or power. One type of biomass is wood, which is frequently used in industry, either by itself to create energy or with other combustible matter (such as coal) to burn and create heat. (Wood has been burned for millennia - as solids.)

Biofuel is considered a means of reducing greenhouse gas emissions and increasing energy security by providing an alternative to fossil fuels. However, In October 2007, Nobel Laureate Paul Crutzen published findings that the release of Nitrous Oxide (N_2O) among the commonly used biofuels, such as biodiesel from rapeseed and bioethanol from corn (maize), can contribute as much or more to global warming than cooling by fossil fuel fuels. Crops with less N demand, such as grasses and woody coppice species have more favourable climate impacts.

Biofuels are used globally: biofuel industries are expanding in Europe, Asia and the Americas. The most common use for biofuels is in automotive transport (for example E10 fuel). Biofuel can be produced from any carbon source that can be replenished rapidly e.g. plants. Many different plants and plant-derived materials are used for biofuel manufacture.

Biomass

Biomass is material derived from recently living organisms. This includes plants, animals and their by-products. For example, manure, garden waste and crop residues are all sources of biomass. It is a renewable energy source based on the carbon cycle, unlike other natural resources such as petroleum, coal, and nuclear fuels. Agricultural products specifically grown for biofuel production include corn, switchgrass, and soybeans, primarily in the United States; rapeseed, wheat and sugar beet primarily in Europe; sugar cane in Brazil; palm oil and miscanthus in South-East Asia; sorghum and cassava in China; and jatropha in India. Hemp has also been proven to work as a biofuel. Biodegradable outputs from industry, agriculture, forestry and households can be used for biofuel production, either using anaerobic digestion to produce biogas, or using second generation biofuel processes; examples include straw, timber, manure, rice husks, sewage, and food waste. The use of biomass fuels can therefore contribute to waste management as well as fuel security and help to prevent climate change, though alone they are not a comprehensive solution to these problems.

History

Humans have used biomass fuels - that is, solid biofuels - for heating and cooking since the discovery of fire. Following the discovery of electricity, it became possible to use biofuels to generate electrical power as well. However, the discovery and use of fossil fuels: coal, gas and oil, have dramatically reduced the amount of biomass fuel used in the developed world for transport, heat and power.

Liquid biofuels have been used since the early days of the car industry. Nikolaus August Otto, the German inventor of the internal combustion engine, conceived his invention to run on ethanol. Rudolf Diesel, the German inventor of the Diesel engine, designed it to run on peanut oil. Henry Ford originally designed the Ford Model T, a car produced from 1903 to 1926, to run completely on ethanol. However, when crude oil became cheaply available (thanks to oil reserves discovered in Pennsylvania and Texas), cars began using fuels derived from mineral oil: petroleum or diesel.

Nevertheless, before World War II, biofuels were seen as providing an alternative to imported oil. Germany powered its vehicles using a blend of gasoline with alcohol fermented from potatoes, called *Reichskraftsprit*. In Britain, grain alcohol was blended with petrol by the Distillers Company Limited under the name *Discol* and marketed through Esso's affiliate Cleveland.

After the war, cheap Middle Eastern oil lessened interest in biofuels. But the oil shocks of 1973 and 1979 increased interest from governments and academics. The counter-shock of 1996 again reduced oil prices and interest.

In the United States, all cars manufactured since 1988 are required to be compatible with fuels containing at least 20% ethanol E20 fuel, and with minor modifications these cars can use 85% ethanol blended with petroleum E85 fuel. Since around 2000 renewed interest in biofuels has been seen. The drivers for biofuel use and development include rising oil prices, concerns over the potential oil peak, greenhouse gas emissions (global warming), rural development interests, and instability in the Middle East. The US president George W. Bush said in his 2006 State of the Union speech that the US should replace 75% of imported oil with biofuel by 2025.

The U.S. Dept. of Energy has earmarked $375 million to fund bioenergy research centers. Second generation biofuel production processes are in development (see below). These allow biofuel to be derived from any source of biomass, not just from food crops such as corn and soy beans.

Carbon Emissions

Biofuels and other forms of renewable energy aim to be carbon neutral. This means that the carbon released during the use of the fuel, e.g. through burning to power transport or generate electricity, is reabsorbed and balanced by the carbon absorbed by new plant growth. These plants are then harvested to make the next batch of fuel. Carbon neutral fuels lead to no net increases in atmospheric carbon dioxide levels, which means that global warming need not get any worse.

In practice, biofuels are not carbon neutral. This is because energy is required to grow crops and process them into fuel. Examples of energy use during the production of biofuels include: fertilizer manufacture, fuel used to power machinery, and fuel used to transport crops and fuels to and from biofuel processing plants. The amount of fuel used during biofuel production has a large impact on the overall greenhouse gas emissions savings achieved by biofuels.

In October 2007, further doubt has been thrown on the advantages of biofuel by Nobel Laureate Paul Crutzen, who says that the advantages of reduced carbon dioxide emissions are more than offset by increased nitrous oxide emissions. Nitrous oxide is both a potent greenhouse gas and a destroyer of atmospheric ozone.

The carbon emissions produced by biofuels are calculated using a technique called Life Cycle Analysis (LCA). This uses a "cradle to grave" or "well to wheels" approach to calculate the total amount of carbon dioxide and other greenhouse gases emitted during biofuel production, from putting seed in the ground to using the fuel in cars and trucks. Many different LCAs have been done for different biofuels, with widely differing results. The majority of LCA studies show that biofuels provide significant greenhouse gas emissions savings when compared to fossil fuels such as petroleum and diesel. Therefore, using biofuels to replace a proportion of the fossil fuels that are burned for transportation can reduce overall greenhouse gas emissions.

This does assume however that the land used for growing the crops would alternatively be desert or paved area. If the land was previously a (tropical rain-) forest, the carbon absorption of this forest should be deducted from the greenhouse gas savings. This implies that the net effect of burning bio-fuels is an increase in greenhouse gasses. This effect should be incorporated in the LCA, to get a proper overview of the total net effect. Using waste material from plantation forests on previous agricultural land could be carbon positive, due to the carbon stored below ground in the root systems.

The well-to-wheel analysis for biofuels has shown that first generation biofuels can save up to 60% carbon emission and second generation biofuels can save up to 80% as opposed to using fossil fuels.

However, a 2007 study by scientists from Britain, U.S., Germany, Switzerland and including Professor Paul Crutzen, who won a Nobel Prize for his work on ozone, have reported that measurements of emissions from the burning of biofuels derived from rapeseed and corn have been found to produce more greenhouse gas emissions than they save.

The claim that biofuels result in emissions savings has also been critiqued on the grounds that it overlooks the 'displacement' effects of large-scale biofuel production, in terms of its direct and indirect role in promoting land use changes and soil carbon losses.

In 2006, a UK Government study showed that carbon emissions were reduced between 50% and 60%. This was when biofuels were used in conjunction with other fuels such as petrol and diesel.

Bioenergy From Waste

Using waste biomass to produce energy can reduce the use of fossil fuels, reduce greenhouse gas emissions and reduce pollution and waste management problems. A recent publication by the European Union highlighted the potential for waste-derived bioenergy to contribute to the reduction of global warming. The report concluded that 19 million tons of oil equivalent is available from biomass by 2020, 46% from bio-wastes: municipal solid waste (MSW), agricultural residues, farm waste and other biodegradable waste streams.

Landfill sites generate gases as the waste buried in them undergoes anaerobic digestion. These gases are known collectively as landfill gas: this can be burned and is a source of renewable energy. Landfill gas (LFG) can be burned either directly for heat or to generate electricity for public consumption. Landfill gas contains approximately 50 percent methane, the same gas that is found in natural gas.

If landfill gas is not harvested, it escapes into the atmosphere: this is not desirable because methane is a greenhouse gas, with more global warming potential than carbon dioxide. Over a time span of 100 years, methane has a global warming potential of 23 relative to CO_2. Therefore, during this time, one ton of methane produces the same greenhouse gas (GHG) effect as 23 tons of CO_2. When methane burns the formula is $CH_4 + 2O_2 = CO_2 + 2H_2O$. So by harvesting and burning landfill gas, its global warming potential is reduced a factor of 23, in addition to providing energy for heat and power.

PhD Frank Keppler and PhD Thomas Rockmann discovered that living plants also produce methane CH_4. [citation needed] The amount of methane produced by living plants is 10 to 100 times greater than that produced by dead plants but does not increase global warming because of the carbon cycle.

Anaerobic digestion can be used as a distinct waste management strategy to reduce the amount of waste sent to landfill and generate methane, or biogas. Any form of biomass can be used in anaerobic digestion and will break down to produce methane, which can be harvested and burned to generate heat, power or to power certain automotive vehicles.

A 3 MW landfill power plant would power 1,900 homes. It would eliminate 6,000 tons per year of methane from getting into the environment. It would eliminate 18,000 tons per year of CO_2 from fossil fuel replacement. This is the same as removing 25,000 cars from the road, or planting 36,000 acres (146 km^2) of forest, or not using 305,000 barrels of oil per year.

First Generation Biofuels

'First-generation fuels' refer to biofuels made from sugar, starch, vegetable oil, or animal fats using conventional technology.

The most common first generation biofuels are listed below.

Vegetable Oil

Vegetable oil can be used for either food or fuel; the quality of the oil may be lower for fuel use. Vegetable oil can be used in many older diesel engines (equipped with indirect injection systems), but only in warm climates. In most cases, vegetable oil is used to manufacture biodiesel, which is compatible with most diesel engines when blended with conventional diesel fuel. MAN B&W Diesel, Wartsila and Deutz AG offer engines that are compatible with straight vegetable oil. Used vegetable oil is increasingly being processed into biodiesel, and at a smaller scale, cleaned of water and particulates and used as a fuel.

Biodiesel

Biodiesel is the most common biofuel in Europe. It is produced from oils or fats using transesterification and is a liquid similar in composition to mineral diesel. Its chemical name is fatty acid methyl (or ethyl) ester (FAME). Oils are mixed with sodium hydroxide and methanol (or ethanol) and the chemical reaction produces biodiesel (FAME) and glycerol. 1 part glycerol is produced for every 10 parts biodiesel.

Biodiesel can be used in any diesel engine when mixed with mineral diesel. In some countries manufacturers cover their diesel engines under warranty for 100% biodiesel use, although Volkswagen Germany, for example, asks drivers to make a telephone check with the VW environmental services department before switching to 100% biodiesel (see biodiesel use). Many people have run their vehicles on biodiesel without problems. However, the majority of vehicle manufacturers limit their recommendations to 15% biodiesel blended with mineral diesel. In many European countries, a 5% biodiesel blend is widely used and is available at thousands of gas stations.

In the USA, more than 80% of commercial trucks and city buses run on diesel. Therefore "the nascent U.S. market for biodiesel is growing at a staggering rate—from 25 million gallons per year in 2004 to 78 million gallons by the beginning of 2005. By the end of 2006 biodiesel production was estimated to increase fourfold to more than 1 billion gallons," energy expert Will Thurmond writes in an article for the July-August 2007 issue of THE FUTURIST magazine.

Bioalcohols

Biologically produced alcohols, most commonly ethanol and less commonly propanol and butanol, are produced by the action of microorganisms and enzymes through fermentation.

Butanol

Butanol is often claimed to provide a direct replacement for gasoline, because it can be used directly in a gasoline engine (in a similar way to biodiesel in diesel engines). It is not in widespread production, and engine manufacturers have not made statements about its use While on paper (and a few lab tests) it appears that butanol has sufficiently similar characteristics with gasoline such that it should work without problem in any gasoline engine, no widespread experience exists. Butanol is formed by ABE fermentation (acetone, butanol, ethanol) and experimental modifications of the process show potentially high net energy gains with butanol as the only liquid product. Butanol will produce more energy and allegedly can be burned "straight" in existing gasoline engines (without modification to the engine or car), and is less corrosive and less water soluble than ethanol, and could be distributed via existing infrastructures. DuPont and BP are working together to help develop Butanol.

Bioethanol

Ethanol is the most common biofuel worldwide. This alcohol fuel is produced by fermentation of sugars derived from wheat, corn, sugar beet and sugar cane. The production methods used are enzymatic digestion (to release sugars from stored starches e.g. from wheat and corn), fermentation of the sugars, distillation and drying. Ethanol can be used in petrol engines as a replacement for gasoline; it can be mixed with gasoline to any percentage, see common ethanol fuel mixtures for information on ethanol. All petrol engines can run on blends of up to 15% bioethanol with petroleum/gasoline. For higher percentage blends, engine modifications are needed. Many car manufacturers are now producing flex-fuel vehicles, which can run on any combination of bioethanol and petrol, up to 100% bioethanol.

Biomethanol

Methanol is currently produced from natural gas, a fossil fuel. It can also be produced from biomass (biomethanol). The methanol economy is an interesting alternative to the hydrogen economy.

BioGas

Biogas is produced by the process of anaerobic digestion of organic material by anaerobes. It can be produced either from biodegradable waste materials or by the use of energy crops fed into anaerobic digesters to supplement gas yields. The solid byproduct, digestate, can be used as a biofuel or a fertiliser.

Biogas contains methane and can be recovered from industrial anaerobic digesters and mechanical biological treatment systems. Landfill gas is a less clean form of biogas which is produced in landfills through naturally occurring anaerobic digestion. If it escapes into the atmosphere it is a potent greenhouse gas.

Oils and gases can be produced from various biological wastes:

- Thermal depolymerization of waste can extract methane and other oils similar to petroleum.
- GreenFuel Technologies Corporation developed a patented bioreactor system that uses nontoxic photosynthetic algae to take in smokestacks flue gases and produce biofuels such as biodiesel, biogas and a dry fuel comparable to coal.

Solid Biofuels

Examples include wood, charcoal, and dried excrement.

Second Generation Biofuels

Second generation biofuels use biomass to liquid technology, including cellulosic biofuels from non food crops.

The following second generation biofuels are under development:

- BioHydrogen
- Bio-DME
- Biomethanol
- DMF

- HTU diesel
- Fischer-Tropsch diesel
- Mixed Alcohols (i.e., mixture of mostly ethanol, propanol and butanol, with some pentanol, hexanol, heptanol and octanol)

Bio-DME, Fischer-Tropsch, BioHydrogen diesel, Biomethanol and Mixed Alcohols all use syngas for production. This syngas is produced by gasification of biomass. HTU (High Temperature Upgrading) diesel is produced from particularly wet biomass stocks using high temperature and pressure to produce an oil.

BioHydrogen is the same as hydrogen except it is produced from a biomass feedstock. This is done using gasification of the biomass and then reforming the methane produced, or alternatively, this can be accomplished with some organisms that produce hydrogen directly under certain conditions. BioHydrogen can be used in fuel cells to produce electricity.

DMF. Recent advances in producing DMF from fructose and glucose using catalytic biomass-to-liquid process have increased its attractiveness.

Bio-DME is the same as DME but is produced from a bio-sources. Bio-DME can be produced from Biomethanol using catalytic dehydration or it can be produced from syngas using DME synthesis. DME can be used in the compression ignition engine.

Biomethanol is the same as methanol but it is produced from biomass. Biomethanol can be blended with petrol up to 10-20% without any infrastructure changes.

HTU diesel is produced from wet biomass. It can be mixed with fossil diesel in any percentage without need for infrastructure.

Fischer-Tropsch diesel (FT)diesel is produced using gas-to-liquids technology. FT diesel can be mixed with fossil diesel at any percentage without need for infrastructure change.

Mixed alcohols are produced from syngas with catalysts similar to those used for methanol. Most R&D in this area is concentrated in producing mostly ethanol. However, some fuels are marketed as mixed alcohols (see Ecalene). Mixed alcohols are superior to pure methanol or ethanol, in that the higher alcohols have higher energy content. Also, when blending, the higher alcohols increase compatibility of gasoline and ethanol, which increases water tolerance and decreases evaporative emissions. In addition, higher alcohols have also lower heat of vaporization than ethanol, which is important for cold starts. (For another method for producing mixed alcohols from biomass see bioconversion of biomass to mixed alcohol fuels)

Wood diesel A new biofuel was developed by the University of Georgia from wood chips. The oil is extracted and then added to unmodified diesel engines. Either new plants are used or planted to replace the old plants. The charcoal byproduct is put back into the soil as a fertilizer. According to the director Tom Adams since carbon is put back into the soil, this biofuel can actually be carbon negative not just carbon neutral. Carbon negative decreases carbon dioxide in the air reversing the greenhouse effect not just reducing it.

Micro Algae

Much research is being done about the use of microalgae as an energy source, with applications for biodiesel, ethanol, methanol, methane and hydrogen. The production of biofuels to replace oil and natural gas is in active development, focusing on the use of cheap organic matter (usually cellulose, agricultural and sewage waste) in the efficient production of liquid and gas biofuels which yield high net energy gain. One advantage of many biofuels over most other fuel types is that they are biodegradable, and so relatively harmless to the environment if spilled.

Biofuels in Developing Countries

Biofuel industries are becoming established in many developing countries. Many developing countries have extensive biomass resources that are becoming more valuable as demand for biomass and biofuels increases. The approaches to biofuel development in different parts of the world varies. Countries such as India and China are developing both bioethanol and biodiesel programs. India is extending plantations of jatropha, an oil-producing tree that is used in biodiesel production. The Indian sugar ethanol program sets a target of 5% bioethanol incorporation into transport fuel Ethanol India website. China is a major bioethanol producer and aims to incorporate 15% bioethanol into transport fuels by 2010.

Amongst rural populations in developing countries, biomass provides the majority of fuel for heat and cooking. Wood, animal dung and crop residues are commonly burned. Figures from the International Energy Agency show that biomass energy provides around 30% of the total primary energy supply in developing countries; over 2 billion people depend on biomass fuels as their primary energy source. world resources institute document on wood fuels

The use of biomass fuels for cooking indoors is a source of health problems and pollution. 1.3 million deaths were attributed to the use of biomass fuels with inadequate ventilation by the International Energy Agency in its World Energy Outlook 2006. Proposed solutions include improved stoves and alternative fuels. However, fuels are easily damaged, and alternative fuels tend to be expensive. People in developing countries are unlikely to be able to afford to put these solutions in place. Organizations such as Intermediate Technology Development Group work to make improved facilities for biofuel use and better alternatives accessible to those who cannot get them.

Efforts and Promotion

Recognizing the importance of implementing bioenergy, there are international organizations such as IEA Bioenergy, established in 1978 by the International Energy Agency (IEA), with the aim of improving cooperation and information exchange between countries that have national programs in bioenergy research, development and deployment.

In Brazil, the government hopes to build on the success of the Proálcool ethanol program by expanding the production of biodiesel which must contain 2% biodiesel by 2008, increasing to 5% by 2013. Colombia mandates the use of 10% ethanol in all gasoline sold in cities with populations exceeding 500,000. In Venezuela, the state oil company is supporting the construction of 15 sugar cane distilleries over the next five years, as the government introduces a E10 (10% ethanol) blending mandate. An EU directive has set the goal of replacing 5.75% of transportation fuel by biofuels by 2010 in all member states. In Canada, the government aims for 45% of the country's gasoline consumption to contain 10% ethanol by 2010. In Southeast Asia, Thailand has mandated an ambitious 10% ethanol mix in gasoline starting in 2007. For similar reasons, the palm oil industry plans to supply an increasing portion of national diesel fuel requirements in Malaysia and Indonesia. In India, a bio-ethanol program calls for E5 blends throughout most of the country targeting to raise this requirement to E10 and then E20. In China, the government is making E10 blends mandatory in five provinces that account for 16% of the nation's passenger cars.

European Union

The European Union has set a goal:

- For 2010 that each member state should achieve at least 5.75% biofuel usage of all used traffic fuel.
- For 2020, 10 % .

USA

A senior member of the House Energy and Commerce Committee Congressman Fred Upton has legislation to use at least E10 fuel by 2012 in all cars in the USA. The fuel is less expensive and is cleaner for the environment. It has higher octane so all cars made after 2012 can be made with a higher compression rating to get better fuel economy.

Oregon

Oregon Governor Ted Kulongoski signed legislation in July 2007 that will require all gasoline sold in the state to be blended with 10% bioethanol (a blend known as BE10) and all diesel fuel sold in the state to be blended with 2% biodiesel (a blend known as BD2).

Current Issues in Biofuel Production and Use

Biofuels can provide benefits including: reduction of greenhouse gas emissions, reduction of fossil fuel use, increased national energy security, increased rural development and a sustainable fuel supply for the future.

However, biofuels have limitations. The feedstocks for biofuel production must be replaced rapidly and biofuel production processes must be designed and implemented so as to supply the maximum amount of fuel at the cheapest cost, while providing maximum environmental benefits. Broadly speaking, first generation biofuel production processes cannot supply us with more than a few percent of our energy requirements sustainably. The reasons for this are described below. Second generation processes can supply us with more biofuel, with better environmental gains. The major barrier to the development of second generation biofuel processes is their capital cost: establishing second generation biodiesel plants has been estimated at •500million Nexant Chem Systems study

Rising Food Prices/the "Food vs. Fuel" Debate

Due to rising demand for biofuels, farmers worldwide have an increased economic incentive to grow crops for biofuel production instead of food production. Without political intervention, this could lead to reduced food production and increased food prices and inflation. The impacts of this would be greatest on poorer countries or countries that rely on imported food for their subsistence.

In early 2007 there were a number of reports linking stories as diverse as food riots in Mexico due to rising prices of corn for tortillas and reduced profits at Heineken, the large international brewer, to the increasing use of corn (maize) grown in the US Midwest for bio-ethanol production. (In the case of beer, the barley area was cut in order to increase corn production. Barley is not currently used to produce bioethanol.)

Environmental campaigner George Monbiot has argued in the British newspaper The Guardian for a 5-year freeze on biofuels while their impact on poor communities and the environment is assessed. One problem with this approach is that economic drivers are required in order to push through the development of more sustainable second generation biofuel processes: these will be stalled if biofuel production decreases. Supporters of biofuels claim that a more viable solution is to increase political and industrial support for, and rapidity of, second generation biofuel implementation from non food crops, including cellulosic biofuels.

The most recent UN report on biofuel also raises issues regarding food security and biofuel production. Jean Ziegler, the UN Special Rapporteur on food, concluded that while the argument for biofuels in terms of energy efficiency and climate change are legitimate, the effects for the world's hungry of transforming wheat and maize crops into biofuel are "absolutely catastrophic," and terms such use of arable land a "crime against humanity." Ziegler also calls for a 5-year moratorium on biofuel production.

Food surpluses exist in many developed countries. For example, the UK wheat surplus was around 2 million tonnes in 2005 (Defra figures after exports,). This surplus alone could produce sufficient bioethanol to replace around 2.5% of the UK's petroleum consumption, without requiring any increase in wheat cultivation or reduction in food supply or exports. However, above a few percent (i.e. if the UK wanted to replace more than around 5% of its fuel with biofuel), there would be direct competition between first generation biofuel production and food production. This is one reason why many view second generation biofuel production processes as increasingly important.

Second generation biofuel production processes use non food crops. These include the stalks of wheat and corn, wood, special energy or biomass crops (e.g. Miscanthus) and waste biomass. These processes could utilise the waste products of current food-based agriculture to manufacture fuel sustainably. Second generation biofuel processes are in development: pilot plants are established for the production of ethanol from wheat straw and of syn-diesel from wood chippings. It is important to note that carbon in waste biomass is used by other organisms, e.g. it is broken down in the soil to produce nutrients, and provides a habitat for wildlife. The large scale use of such "waste" biomass by humans might threaten these habitats and organisms.

Biofuel Prices

Retail, at the pump prices, including Federal and state motor taxes, B2/B5 prices for low-level Biodiesel (B2-B5) are lower than petroleum diesel by about 12 cents, and B20 blends are the same as petrodiesel. [27]

Poverty Reduction

Researchers at the Overseas Development Institute have argued that biofuels could help to reduce poverty in the developing world, through increased employment, wider economic growth multipliers and energy price effects. However, this potential is described as 'fragile', and is reduced where feedstock production tends to be large scale, or causes pressure on land access. With regards to potential for poverty reduction, biofuels rely on many of the same policy, regulatory or investment shortcomings that impede agriculture as a route to poverty reduction. As many of these shortcomings require policy improvements at a country level, rather than a global one, they argue for a country-by-country analysis of the potential poverty impacts of biofuels. This would consider, among other things, land administration systems, market coordination and prioritising investment in biodiesel as this 'generates more labour, has lower transportation costs and uses simpler technology'.

Energy Efficiency and Energy Balance of Biofuels

Production of biofuels from raw materials requires energy (for farming, transport and conversion to final product as well as the production of fertilizers, pesticides and herbicides). The level of energy expenditure varies by location: more intensive agricultural regimes such as those found in Western countries are more energy intensive. The more machinery is used for farming, the greater the energy expended in the process; developing countries tend to have less intensive agricultural methods. It is possible to produce biomass without incurring large agricultural energy costs: for example, wild-harvesting excess wood from established forests can be done without much energy input. However the yield of biomass from such resources is not consistent or large enough to support biofuel manufacture on a large scale.

The energy balance of a biofuel is determined by the amount of energy put into the manufacture of fuel compared to the amount of energy released when it is burned in a vehicle and some biofuels can produce up to 2-36 times the input rate of fossil fuels. Biofuels tend to require higher energy inputs per unit energy than fossil fuels: oil can be pumped out of the ground and processed more efficiently than biofuels can be grown and processed. However, this is not necessarily a reason to use oil instead of biofuels, nor does it have an impact on the environmental benefits provided by a given biofuel.

Other factors connected to energy balance are a) cost and b) environmental impact. High energy impacts do not necessarily mean that the resulting fuel will be bad for the environment: energy can be derived from renewable resources to power biofuel manufacture.

Energy balance is not necessarily a measure of a good biofuel. Biofuels should be affordable, sustainable, abundant and provide good GHG emissions savings when compared with fossil fuels.

Energy balance/ efficiency of conversion is relevant when considering how best to use a given amount of biomass resources. For example, given limited resources should biomass be converted into heat and power or liquid transport fuels? Looking at energy balance and the efficiency of energy conversion can help to use biomass resources efficiently and with maximum environmental gain.

Studies have been done that calculate energy balances for biofuel production. Some of these show large differences depending on the biomass feedstock used and location.

The energy balance is more favourable for biofuels made from crops grown in subtropical or tropical areas than those made from crops grown in temperate areas. This is largely due to the increased yield of biomass from crops in areas that receive more sunlight.

Life cycle assessments of biofuel production show that under certain circumstances, biofuels produce only limited savings in energy and greenhouse gas emissions. Fertiliser inputs and transportation of biomass across large distances can reduce the GHG savings achieved. The location of biofuel processing plants can be planned to minimize the need for transport, and agricultural regimes can be developed to limit the amount of fertiliser used for biomass production. A European study on the greenhouse gas emissions found that well-to-wheel (WTW) CO_2 emissions of biodiesel from seed crops such as rapeseed could be almost as high as fossil diesel. It showed a similar result for bio-ethanol from starch crops, which could have almost as many WTW CO_2 emissions as fossil petrol. This study showed that second generation biofuels have far lower WTW CO_2 emissions.

Other independent LCA studies show that biofuels save around 50% of the CO_2 emissions of the equivalent fossil fuels. This can be increased to 80-90% GHG emissions savings if second generation processes or reduced fertiliser growing regimes are used (Concawe Well to Wheels LCA for biofuels).

Environmental Effects

Some mainstream environmental groups support biofuels as a significant step toward slowing or stopping global climate change. However, biofuel production can threaten the environment if it is not done sustainably. This finding has been backed by reports of the UN, the IPCC and some other smaller environmental and social groups as the EEB and the Bank Sarasin ,which generally remain negative about biofuels.

As a result, governmental and environmental organisations are turning against biofuels made at a non-sustainable way (hereby preferring certain oil sources as jatropha and lignocellulose over palm oil) and are asking for global support for this

Also, besides supporting these more sustainable biofuels, environmental organisations are redirecting to new technologies that do not use combustion engines as hydrogen and compressed air.

Biofuels produce greenhouse gas emissions during their manufacture. The source of these emissions are: fertilisers and agricultural processing, transportation of the biomass, processing of the fuels, and transport and delivery of biofuels to the consumer. Some biofuel production processes produce far fewer emissions than others; for example sugar cane cultivation requires fewer fertiliser inputs than corn cultivation, therefore sugar cane bioethanol reduces greenhouse gas emissions more effectively than corn derived bioethanol. However, given the appropriate agricultural techniques and processing strategies, biofuels can provide emissions savings of at least 50% when compared to fossil fuels such as diesel and petroleum.

The increased manufacture of biofuels will require increasing land areas to be used for agriculture. Second generation biofuel processes can ease the pressure on land, because they can use waste biomass, and existing (untapped) sources of biomass such as crop residues and potentially even marine algae.

In some regions of the world, a combination of increasing demand for food, and increasing demand for biofuel, is causing deforestation and threats to biodiversity. The best reported example of this is the expansion of oil palm plantations in Malaysia and Indonesia, where rainforest is being destroyed to establish new oil palm plantations. It is an important fact that 90% of the palm oil produced in Malaysia is used by the food industry Malaysian Palm Oil Council; therefore biofuels cannot be held solely responsible for this deforestation. There is a pressing need for sustainable palm oil production for the food and fuel industries; palm oil is used in a wide variety of food products. The Roundtable on Sustainable Biofuels is working to define criteria, standards and processes to promote sustainably produced biofuels. Palm oil is also used in the manufacture of detergents, and in electricity and heat generation both in Asia and around the world (the UK burns palm oil in coal-fired power stations to generate electricity).

Significant area is likely to be dedicated to sugar cane in future years as demand for ethanol increases worldwide. The expansion of sugar cane plantations will place pressure on environmentally-sensitive native ecosystems including rainforest in South America. In forest ecosystems, these effects themselves will undermine the climate benefits of alternative fuels, in addition to representing a major threat to global biodiversity.

Although biofuels are generally considered to improve net carbon output, biodiesel and other fuels do produce local air pollution, including nitrogen oxides, the principal cause of smog.

COMPOSTING

Composting can be divided into home composting and industrial composting. Both scales of composting use the same biological processes, however techniques and different factors must be taken into account.

Importance

Composting recycles organic household and yard waste into compost, which returns needed organic matter to the soil. While improving the soil for gardens, composting also reduces the amount of waste going into burgeoning landfills. Composting speeds up the process of decompositon by providing an optimal environment for waste to break down and enable it to return to the earth. This requires the correct mix of carbon, nitrogen, oxygen and water.

Decomposition occurs even without some of these ingredients; however, the process can be slow and unpleasant. For example, vegetables in a plastic bag decompose, but the lack of air encourages the growth of anaerobic microbes, which produce disagreeable odors. Degradation with insufficient air is called anaerobic digestion.

Materials

Different materials contain various nutrients for the composting bacteria. Some materials contain much carbon in the form of cellulose which the bacteria need to produce sugars and heat. Other materials contain nitrogen in the form of protein, which allow the compost bacteria to thrive.

Ingredients with higher carbon content include:

- Dry, straw-type material, such as cereal straws
- Autumn leaves
- Sawdust and wood chips
- Some paper and cardboard (such as corrugated cardboard or newsprint with soy-based inks)

Ingredients with higher nitrogen content include:

- Green plant material (fresh or wilted) such as crop residues, hay, grass clippings, weeds
- Manure of poultry and herbivorous animals such as horses, cows and llamas
- Fruit and vegetable trimmings

- Seaweeds
- Coffee grounds

The most rapid composting occurs in matter containing 25 to 30 times more carbon than nitrogen by dry chemical weight. Grass clippings have an average ratio of 19 to 1 and dry autumn leaves about 55 to 1. Mixing equal parts by volume approximates the ideal range.

Poultry manure provides much nitrogen but little carbon, while horse manure provides both. Sheep and cattle manure compost at a lower temperature and so slower than poultry or horse manure.

Eggshells are a good source of nutrients for the compost pile and the soil, but typically take more than one year to decompose.

Mixing the materials as they are added increases the rate of decomposition, but it can be easier to place the materials in alternating layers, approximately 15 cm (6 in) thick, to help estimate the quantities. Keeping carbon and nitrogen sources separated in the pile can slow down the process, but decomposition will occur.

Some people put special materials and activators into their compost. A light dusting of agricultural lime (not on animal manure layers) can curb excessive acidity. Seaweed meal provides a ready source of trace elements. Finely pulverized rock (rock flour or rock dust) can also provide minerals, while clay and leached rock dust are poor in trace minerals.

Composting also breaks down petroleum hydrocarbons and some toxic compounds for recycling and beneficial reuse. Composting for such purposes is called bioremediation.

Some materials are best left to high-rate, a thermophilic composting system, as they decompose slower, attract vermin and require higher temperatures to kill pathogens than backyard composting provides. These materials include meat, dairy products, eggs, restaurant grease, cooking oil, manure and bedding of non-herbivores, and residuals from the treatment of wastewater and drinking water. Meat and dairy products can be recycled using bokashi, a fermentation method, but milk and oil cannot.

Human waste can be composted by composting toilets, but most composting toilets do not allow for the thermophilic decomposition needed to kill pathogens. If these high temperatures are reached, the resulting compost can be safely used as a fertilizer for food crops and even directly edible crops. Careful filtration of the compost also prevents contamination.

Approaches

There are two major approaches to composting: active and passive.

Active

Active (hot) composting is composting at close to ideal conditions, allowing aerobic bacteria to thrive. Aerobic bacteria break down material faster and produce less odor and fewer pathogens and destructive greenhouse gases than anaerobic bacteria. Commercial-grade composting operations actively control the composting conditions such as the carbon-to-nitrogen ratio. For backyard composters, the charts of carbon and nitrogen ratios in various ingredients and the calculations required to get the ideal mixture can be intimidating, so many rules of thumb exist for approximating it.

Pasteurisation in a hot compost (such as a compost oven) occurs if the temperature exceeds 55 °C (130 °F) for three or more days. To achieve it, the compost bin must be kept warm, insulated and damp to encourage the cultivation of actinomycete bacteria. Pasteurised soil is valuable for gardening; pasteurisation is otherwise expensive and complicated, and adding chemicals to produce it makes the compost less healthy.

Aerated Composting is a particularly efficient form of hot composting. With aerated composting, fresh air (i.e. oxygen) is introduced throughout the mix of materials using an electric blower. The oxygen stimulates the micro-organisms that are already in the mix, and their by-product is heat. In a properly operated compost system, pile temperatures are sufficient to pasteurize the raw material, and the oxygen-rich conditions within the core of the pile eliminate offensive odors. High temperatures also destroy fly larvae and weed seeds, yielding a safe, high-quality finished product.

Finally, aeration greatly expedites the composting process and a finished product can be rendered in 30 to 60 days, without turning the compost pile. Aerated compost is an excellent source of macro- and micro-nutrients as well as stable organic matter, all of which support healthy plant growth. In addition, the micro-organisms in compost aid in the suppression of plant pathogens. Finally, compost retains water extremely well resulting in improved drought resistance, a longer growing season, and reduced soil erosion.

Civil engineer and environmental consultant Peter Moon is a pioneer in the field of aerated composting. He has provided solutions to waste management problems for both large scale agricultural industries and smaller private enterprises such as horse ranches Learn more about aerated composting at http://o2compost.com/index.htm.

Passive

Passive (cold) composting is composting in which temperatures never reach above 30 °C (86 °F). It is slower and leaves dormant pathogens, but is the more common type of composting in most domestic garden compost bins. Such composting systems may be either enclosed (home container composting, industrial in-vessel composting) or in exposed piles (industrial windrow composting). Kitchen scraps are put in the garden compost bin and left untended. This scrap bin has a very high water content and no aeration, and so becomes anaerobic and creates foul smell and emits significant adverse greenhouse gases. To improve drainage and airflow, a gardener can mix in wood chips, small pieces of bark, leaves or twigs.

Home Composting

Home composters use a range of techniques, varying from extremely passive (throw everything in a pile and leave it for a year or two) to extremely active (monitor the temperature, turn the pile regularly, and adjust the ingredients over time). Some composters use mineral powders to absorb smells, although a well-maintained pile seldom has bad odors.

Moisture and Heat

An effective compost pile is about as damp as a well wrung-out sponge. This provides the moisture that all life requires. Microorganisms vary by their ideal temperature and the heat they generate as they digest. Mesophilic bacteria enjoy temperatures of 20 to 40 °C (70 to 110 °F). thermophilic (heat-loving) bacteria prefer around 60 °C (140 °F), and are the fastest decomposers. This is the ideal temperature as it also kills most pathogens and weed seeds. To achieve it, the heap should be about 1 m (3 ft) wide, 1 m (3 ft) tall, and as long as is practicable. This provides enough insulating mass to build up heat but also allows aeration. The center of the pile heats up the most.

If the pile does not heat up, common reasons include that:

- The heap is too wet, limiting the oxygen which bacteria require
- The heap is too dry for the bacteria to survive and reproduce
- There is insufficient protein (nitrogen-rich material)

The necessary material should be added, or the pile should be turned to aerate it and bring the outer layers inside and vice versa. Adding water at this time helps keep the pile damp. One guideline is to turn the pile when the high temperature has begun to drop, indicating that the food source for the fastest-acting bacteria (in the center of the pile) has been largely consumed. When turning the pile does not cause a temperature rise, it brings no further advantage. When all the material has turned into dark brown crumbly matter, it is ready to use.

Worm Composting

Worm composting or vermicomposting is a method of composting using Red Wiggler worms in a container. Food waste and moistened bedding are added, and the worms and micro-organisms eventually convert them to rich compost. The worms excrete a soil-nutrient material called worm castings. Worm composting can be done indoors, allowing year-round composting, and providing apartment dwellers with a means of composting.

Worms are low in the food chain, and so are critical to healthy soil. This is why farmers have historically wanted healthy worm populations to live in their fields.

The nutrients and microorganisms can be concentrated in liquid form called worm tea, made by running distilled water through worm castings. When it is poured into the soil, the microorganisms multiply, creating a healthy growing environment for plants.

Industrial Composting

Industrial composting systems are increasingly being installed as a waste management alternative to landfills, along with other advanced waste processing systems. It has been done even for landfills from foundry furan resin sand dust. Industrial composting or anaerobic digestion combined with mechanical sorting of mixed waste streams and is called mechanical biological treatment. Treating biodegradable waste before it enters a landfill reduces global warming; untreated waste breaks down anaerobically in a landfill, producing landfill gas that contains methane, a potent greenhouse gas.

Most commercial and industrial composting operations use active composting techniques. This ensures a high-quality product and a short processing time (see compost windrow turner). The greatest control, and so the highest quality, is achieved by composting inside an enclosed vessel and controlling its temperature, air flow, moisture and other parameters. See In-vessel composting (indoor composting) (also called en-vessel).

Large-scale composting systems are used by a few urban centers around the world. Co-composting is a technique which combines solid waste with de-watered biosolids. The world's largest co-composter is the Edmonton Composting Facility in Edmonton in Alberta, Canada, which turns 220,000 tonnes of residential solid waste and 22,500 dry tonnes of biosolids per year into 80,000 tonnes of compost. The facility is 38,690 square metres (416,500 ft²) large (equivalent to 4½ Canadian football fields), and the aeration building alone is the largest stainless steel building in North America, the size of 14 NHL rinks.

REFERENCES

B.C. Patten (1959) An Introduction to the Cybernetics of the Ecosystem: The Trophic-Dynamic Aspect, *Ecology*, Vol. 40, No. 2, pp.221-231.

B.C. Patten and E.P.Odum (1981) The Cybernetic Nature of Ecosystems, *Am. Nat.*, Vol. 118, pp. 886-895.

B.D. Fath, B.C.Patten, and J.S.Choi (2001) Complementarity of ecological goal functions', *Journal of Theoretical Biology*, Volume 208, Issue 4, pp. 493-506.

B.R. Bakshi (2000) 'A thermodynamic framework for ecologically conscious process systems engineering', Computers and Chemical Engineering 24, pp. 1767-1773.

Bilgen, S. and K. Kaygusuz, *Renewable Energy for a Clean and Sustainable Future*, Energy Sources 26, 1119 (2004).

C. Giannantoni (2000) 'Toward a Mathematical Formulation of the Maximum Em-Power Principle', in M.T.Brown (ed.) *Emergy Synthesis: Theory and applications of the emergy methodology*, Proceedings from the first biennial emergy analysis research conference, The Center for Environmental Policy, Department of Environmental Engineering Sciences, University of Florida, Gainesville, FL.

C. Giannantoni (2002) *The Maximum Em-Power Principle as the basis for Thermodynamics of Quality*, Servizi Grafici Editoriali, Padova.

C. Giannantoni (2006) 'Mathematics for generative processes: Living and non-living systems' Journal of Computational and Applied Mathematics 189, pp. 324–340.

D.E. Cambell, S.L. Brandt-Williams (2005) Environmental Accounting Using Emergy: Evaluation of the State of West Virginia, USEPA Office of Research and Development, National Health and Environmental Effects Research Laboratory, Atlantic Ecology Division, Narragansett, RI 02882.

D.E.Campbell (2001) Proposal for including what is valuable to ecosystems in environmental assessments', *Environmental Science and Technology*, Volume 35, Issue 14, pp. 2867-2873.

D.M. Scienceman (1987) 'Energy and Emergy.' In G. Pillet and T. Murota (eds), *Environmental Economics: The Analysis of a Major Interface.* Geneva: R. Leimgruber. pp. 257-276. (CFW-86-26)

D.M. Scienceman (1989) ' The Emergence of Emonomics'. In *Proceedings of The International Society for General Systems Research Conference* (July 2-7, 1989), Edinbrough, Scotland, 7 pp. (CFW-89-02).

D.M. Scienceman (1991) *Emergy and Energy: The Form and Content of Ergon.* Discussion paper. Gainesville: Center for Wetlands, University of Florida. 13 pp. (CFW-91-10)

D.M. Scienceman (1992) *Emvalue and Lavalue*, Paper Prepared for th Annual Meeting of The International Society for the Systems Sciences, University of Denver, Denver, Colorado, U.S.A.

D.M. Scienceman (1997) 'Letters to the Editor: Emergy definition', *Ecological Engineering*, 9, pp. 209-212.

D.M. Scienceman 1995, "The Emergy Synthesis of Religion and Science", Center for Environmental Policy, University of Florida.

D.M. Scienceman and B.M. El-Youssef (1993) The System of Emergy Units, in Packham, R. ed. *Ethical management of science as a system*, International Society for the Systems Sciences, proceedings of the thrity-seventh annual meeting, University of Western Sydney, Hawkwsbury, July 5-9, pp 214-223.

E. Sciubba, S. Ulgiatib (2005) 'Emergy and exergy analyses: Complementary methods or irreducible ideological options?' Energy 30, pp. 1953–1988.

E.N. Transeau (1926) 'The accumulation of energy by plants', *Ohio Journal of Science*, 26, 1-10.

E.P. Odum (1962) 'Relationships between structure and function in ecosystems, *Japanese Journal of Ecology*, 12, 108-118.

Energy analysis of Power Systems, UIC Nuclear Issues Briefing Paper 57 (2004).

G.P. Genoni (1997) 'Towards a conceptual synthesis in ecotoxicology', *OIKOS*, 80:1, pp. 96-106.

G.P. Genoni, E.I. Meyer and A.Ulrich (2003) 'Energy flow and elemental concentrations in the Steina River ecosystem (Black Forest,Germany)', *Aquat. Sci.*, Vol. 65, pp. 143–157.

G.Q. Chen (2006) 'Scarcity of exergy and ecological evaluation based on embodied exergy', *Communications in Nonlinear Science and Numerical Simulation*, 11, pp. 531–552

H.T. Odum (1970) Energy Values of Water Sources. in 19th Southern Water Resources and Pollution Control Conference.

H.T. Odum (1984) Embodied Energy, Foreign Trade, and Welfare of Nations, in A-M. Jansson (ed.) *Integration of Economy and Ecology - An Outlook for the Eighties.*

H.T. Odum (1986) in N.Polunin, Ed. *Ecosystem Theory and Application*, Wiley, New York.

H.T. Odum (1988) 'Self-Organization, Transformity, and Information', *Science*, Vol. 242, pp. 1132-1139.

H.T. Odum (1995) 'Self-Organization and Maximum Empower', in C.A.S. Hall (ed.) *Maximum Power; The Ideas and Applications of H.T.Odum*, Colorado University Press, Colorado, pp. 311-330.

H.T. Odum (1995b) Emergy: Policies for a New World Order, in H.Abele (ed.) *Energy and Environment; A question of Survival*, Verlag Stiftsdruckerei, Switzerland.

H.T. Odum (1996) *Environmental Accounting: Emergy and Environmental Decision Making*, Wiley.

H.T. Odum (1997) Letter to the Editor, *Ecological Engineering*, 9, 215-216.

H.T. Odum (2002) 'Material circulation, energy hierarchy, and building construction', in C.J.Kibert, J.Sendzimir, and G.B.Guy (eds) *Construction Ecology; Nature as the basis for green buildings*, Spon Press, New York.

H.T. Odum and E.C.Odum (1983)*Energy Analysis Overview of Nations*, Working Paper, WP-83-82. Laxenburg, Austria: International Institute of Applied System Analysis. 469 pp. (CFW-83-21).

H.T. Odum and E.C.Odum (2001) *A Prosperous way Down: Principles and Policies*, Colorado University Press, Colorado.

J. Laganisa, & M. Debeljakb (2006) 'Sensitivity analysis of the emergy flows at the solar salt production process in Slovenia', *Journal of Ecological Modelling*, 194, pp. 287–295.

J.L. Hau and B.R.Bakshi (2004) Promise and Problems of Emergy Analysis', *Ecological Modelling*, special issue in honor of H. T. Odum, vol. 178, pp. 215-225.

J.R. Richardson (1988) *Spatial patterns and maximum power in ecosystems*, Ph.D. Dissertation, University of Florida.

M.T. Brown (1997) Letters to the editor, *Ecological Engineering*, 9, 213-214.

M.T. Brown and S. Ulgiati (2004) Energy quality, emergy, and transformity: H.T. Odum's contributions to quantifying and understanding systems, *Ecological Modelling*, Vol. 178, pp. 201-213.

P.K. Nag (1984) *Engineering Thermodynamics*, Tata McGraw-Hill Publishing Company.

S. Bastianoni (2000) 'The problem of co-production in environmental accounting by emergy analysis', Ecological Modelling 129, pp.187–193.

S. Bastianoni, A. Facchini, L. Susani, E. Tiezzi (2007) 'Emergy as a function of exergy', *Energy* 32, 1158-1162.

S. Bastianoni, F.M. Pulselli, M.Rustici (2006) Exergy versus emergy flow in ecosystems: Is there an order in maximizations?', Ecological Indicators 6, pp. 58–62

S. Ulgiati and M.T. Brown (1999) Emergy evaluation of natural capital and biosphere services.

S. Ulgiati and M.T.Brown (2001) 'Emergy Accounting of Human-Dominated, Large-Scale Ecosystems', in S.E.Jorgensen (ed) *Thermodynamics and Ecological Modelling*, CRC Press LLC, pp. 63-113.

S. Ulgiati, H.T. Odum, S.Bastianoni (1994) 'Emergy use, environmental loading and sustainability. An emergy analysis of Italy', *Ecological Modelling*, Volume 73, Issue 3-4, Pages 215-268.

S.E. Jorgensen, S.N.Nielsen, H.Mejer (1995) 'Emergy, environ, exergy and ecological modelling', *Ecological Modelling*, 77, pp. 99-109.

S.E. Tennenbaum (1988) *Network Energy Expenditures for Subsystem Production*, MS Thesis. Gainesville, FL: University of FL, 131 pp. (CFW-88-08)

Serra, J. "Alternative Fuel Resource Development", Clean and Green Fuels Fund, (2006).

Shu-Li Huang and Chia-Wen Chen (2005) 'Theory of urban energetics and mechanisms of urban development', *Ecological Modelling*, 189, pp. 49–71.

T.T. Cai, T.W.Olsen and D.E.Campbell (2004) Maximum (em)power: A foundational principle linking man and nature', *Ecological Modelling*, Volume 178, Issue 1-2, pp. 115-119.

Wang, F.C., Odum, H.T. and Costanza, R. (1980) Energy Criteria for Water Use, *Journal of the Water Resources Planning and Management Division*.

Wang, F.C., Odum, H.T. and Kangas, P.C. (1980) Energy Analysis for Environmental Impact Assessment, *Journal of the Water Resources Planning and Management Division*.

5

Green Building, Green Computing & Other Technologies

NATURAL BUILDING

Natural building involves a range of building systems and materials that place major emphasis on sustainability. Ways of achieving sustainability through natural building focus on durability and the use of minimally-processed, plentiful or renewable resources, as well as those which, while recycled or salvaged, produce healthy living environments and maintain indoor air quality. Natural building tends to rely on human labor, more than technology. As Michael G. Smith observes, it depends on "local ecology, geology and climate; on the character of the particular building site, and on the needs and personalities of the builders and users."

The basis of natural building is the need to lessen the environmental impact of buildings and other supporting systems, without sacrificing comfort, health or aesthetics.[2] To be more sustainable, natural building uses primarily abundantly-available, renewable, reused or recycled materials. The use of rapidly renewable materials is increasingly a focus. An emphasis on building compactly and minimizing the ecological footprint is common, as are on-site handling of energy acquisition, on-site water capture, alternate sewage treatment and water reuse.

Materials

The materials common to many types of natural building are clay and sand. When mixed with water and, usually, straw or another fiber, the mixture may form *cob* or *adobe* (clay blocks). Other materials commonly used in natural building are: earth (as rammed earth or earth bag), wood (cordwood or timber frame/post-and-beam), straw, rice-hulls, bamboo and rock. A wide variety of reused or recycled materials are common in natural building, including urbanite (salvaged chunks of used concrete), tires, tire bales, discarded bottles and other recycled glass.

Several other materials are increasingly avoided by many practitioners of this building approach, due to their major negative environmental or health impacts. These include unsustainably-harvested wood, toxic wood-preservatives, portland cement-based mixes, paints and other coatings which off-gas volatile organic compounds (VOCs), and some plastics, particularly polyvinyl chloride (PVC or "vinyl") and those containing harmful plasticizers or hormone-mimicking formulations.

Techniques

Many traditional methods, techniques, and materials, are now experiencing a resurgence of popularity, however the relative popularity of these techniques differs around the World.

Adobe

One of the oldest building methods, adobe is simply clay and sand mixed with water. Sometimes chopped straw or other fibers are added for strength. The mixture is then allowed to dry in the desired shape. Usually adobe is shaped into bricks that can be stacked to form walls.

Various claims are made about the optimal proportions of clay and sand (or larger aggregate). Some say that the best adobe soil contains 15% - 30% clay to bind the material together. Others say equal proportions of clay and sand are best to prevent cracking or fragmenting of the bricks. Sometimes adobe is stabilized with a small amount of cement or asphalt emulsion to provide better weatherproofing. The blocks can either be poured into molds and dried, or pressed into blocks. Adobe colored with clay and polished with natural oil makes an attractive and resilient floor.

To protect the walls and reduce maintenance, adobe buildings usually have large overhanging eaves and sizeable foundations. Adobe can be plastered over with cob or lime-based mixes for both appearance and protection. Adobe has good thermal mass, meaning that it is slow to transmit heat or cold. It is not a good insulator, however, so insulation can be added (preferably on the outside), or a double wall built with airspace or insulation in between. The traditional thick, un-insulated adobe has proven to perform best in regions without harsh winters or where daily sun is predictably available during those cold periods.

Cob

The term *cob* is used to describe a monolithic building system based on a mixture of clay, sand and straw. The construction uses no forms, bricks or wooden framework; it is built from the ground up. Cob building began use in England prior to the 13th century. Its use spread to other countries with the world-wide dispersal of English colonists, and fell out of favor after World War I, although it is seeing a resurgence today. Similar forms of "mud" building have been used in many parts of the world for centuries, under a variety of names, and date from at least 10,000 years ago. Cob is one of the simplest and least expensive building techniques available, though it is typically very labor-intensive. Cob's other great advantage is versatility; It can easily be shaped into any form. While cob building was falling out of favor in England by the late 19th century, thousands of cob structures have endured to the present (20,000 in Devon, England alone). It is estimated that from one third to one half of the world's population lives in earthen dwellings today. Although typically associated with "low-rise" structures, in Yemen and other Middle-Eastern countries, it has, for centuries, been used in "apartment" buildings of eight stories and more.

Cob-like mixes are also used as plaster or filler in several methods of natural building, such as adobe, earth bags, timber frames, cordwood, and straw bales. Earth is thus a primary ingredient of natural building.

Subterranean Construction

((The most natural constructions in the World are the subterranean houses called silos de Villacañas. With only excavate on land his rooms have windows up.)) ((Tunels, Subways, Nuclear Bunker))

Cordwood

Cordwood construction is a term used for a natural building method in which "cordwood" or short lengths pieces of debarked tree are laid up crosswise with masonry or cob mixtures to build a wall. The cordwood, thus, becomes infill for the walls, usually between posts in a timber frame structure. Cordwood masonry can be combined with other methods (e.g., rammed earth, cob or light clay) to produce attractive combinations. Cordwood masonry construction provides a relatively high thermal mass, which makes it easy to heat and cool.

Earth Bag

Earth is the most typical fill material used in bag-wall construction techniques. This building method utilizes stacked polypropylene or natural-fiber (burlap) bags filled with earth or other mixes, with or without a stabilizer such as portland cement, to form footings, foundations, walls and even vaulted or domed roofs. In recent years, building with earth bags has become one of the increasingly-practiced techniques in natural building. It facilitates self-contained, often free-form rammed-earth structures. Its growing popularity relates to its use of an abundant and readily available often site-available material (earth) in a potentially inexpensive building technique that is flexible, and easy to learn and use. However, because earth is a poor insulator, in more extreme climates other filler variations are now being explored, substituting pumice, rice-hulls or another material with better insulating value for all or part of the earth (*see also* Rice-hull bagwall construction)

Rammed Earth

Rammed earth is an earth-based wall system made of compacted gravel, sand, and clay; that is extremely strong and durable. Quality rammed earth walls are dense, solid, and stone-like with great environmental benefits and superior low maintenance characteristics. As an option depending on climate or seismic concerns rigid insulation can be placed inside the wall as well as steel reinforcement. A truly beautiful material with a spectrum of color choices and textures the walls evoke feelings of security and strength. Used for around 10,000 years in all types of buildings from low rise to high-rise and from small huts to palaces rammed earth has a proven track record.

Rammed earth walls are formed in place by pounding damp sub-soil (gravel, sand, and clay) into movable, reusable forms with manual or machine-powered tampers. A mixture of around 70% aggregate (gravel,sand) and 30% clay is optimal. Cement may be added if the mix requires it or pigmentation to achieve the desired color. Around 5-10 inches of mixed damp sub-soil are placed inside the forms and pounded to total compaction and the process is repeated until the desired height is achieved. What is left after the forms are removed is a beautiful environmental wall that is structural and can last over 1000 years.

Straw Bale

Although grasses and straw have been in use in a range of ways in building since pre-history around the world, their incorporation in machine-manufactured modular bales seems to date back to the early 20th century in the midwestern United States, particularly the sand-hills of Nebraska, where grass was plentiful and other building materials (even quality sods) were not. Straw bale building typically consists of stacking a series of rows of bales (often in running-bond) on a raised footing or foundation, with a moisture barrier between. Bale walls are often tied together with pins of bamboo, rebar, or wood (internal to the bales or on their faces), or with surface wire meshes, and then stuccoed or plastered, either with cementaceous mixes, lime-based formulations or earth/clay renders. Bale buildings can either have a structural frame of other materials, with bales between (simply serving as insulation and stucco substrate), referred to as "infill",or the bales may actually provide the support for openings and roof, referred to as "load-bearing" or "Nebraska-style", or a combination of framing and load-bearing may be employed, referred to a "hybrid" straw bale.

Typically, bales created on farms with mobile machinery have been used ("field-bales"), but recently higher-density "recompressed" bales (or "straw-blocks") are increasing the loads that may be supported; where field bales might support around 600 pounds per linear foot of wall, the high density bales bear up to 4,000 lb./lin.ft. and more. And the basic bale-building method is now increasingly being extended to bound modules of other often-recycled materials, including tire-bales, as well as those of cardboard, paper, plastics and used carpeting, and to bag-contained "bales" of wood-chips, rice-hulls, etc.

Timber Frame

The essential elements of timber frame building—joined timbers, clay walls and thatch roofs were in place in Europe and Asia by the 9th century. It remained the common mode of house construction in northern cultures until the 19th century. Craftsmanship was, and is, an important value in timber frame building. The oldest timber frame structures (for example, the timber framed stave churches of Scandinavia) show both craftsmanship and a strong grasp of the technical aspects of structural design, as do such structures in Japan.

Timber framing typically uses a "bent." A bent is a structural support, like a truss, consisting of two posts, a tie beam and two rafters. These are connected into a framework through joinery. To practice the craft, one must understand the basic structural aspects of the bent. This, along with a knowledge of joinery, are the basis of timber frame building.

Timber framing is now a Modern Method of Construction, Ideally suited to mass house building as well as public buildings. In conjunction with a number of natural insulations and timber cladding or modern lime renders, it is possible to quickly construct a high performance, sustainable building, using completely natural products. The benefits are many - the building will perform better over its lifespan, waste is reduced (much can be re-cycled, composted or used as fuel).

Related Ideas and Strategies

Other concepts, methods and strategies often (or sometimes) associated with natural building include: building "underground," earth sheltering, or berming, "green" or "living" planted roofs, thatched roofs and cement-free earthen floors, rubble-trench, or gabion foundations.

New method for do buildings with adobe doed ceramic in situ, thanks to portable electric horno guided by a machine.

To increase sustainability, various approaches to lower energy consumption are used in conjunction with natural building: sun-shading or other passive cooling techniques, passive solar heating, geo-exchange heating and cooling, "short-cycle" and "annualized" passive (and PV-assisted) solar space and water heating, hot water heat recycling, biologic air purification by indoor plants, passive or air-to-air/heat-recovery ventilation, solar or annualized cooling, insulated glazing and selective glazing films, night and cold-weather "movable" insulation, or on-site electric power generation by renewable energy in the form of photovoltaics (PV), wind generators, or micro-hydro (either with fully-independent systems referred to as "off-grid" or with "grid-tied" systems feeding into the public electric network), low-voltage electric and avoidance of electro-magnetic and other possibly-harmful forms of radiation.

Other green building strategies that improve conservation of resources include: rain-water catchment, storage, and purification; waste-water separation; biological waste-water purification and grey-water reuse; composting toilets, on-site snow/rain-water run-off management; permeable paving; native or low-water-use ("xeriscape") landscapes, and accommodation of alternative-fuelled/powered and human-powered vehicles.

GREEN BUILDING

Green building is the practice of increasing the efficiency of buildings and their use of energy, water, and materials, and reducing building impacts on human health and the environment, through better siting, design, construction, operation, maintenance, and removal — the complete building life cycle.

A similar concept is natural building, which is usually on a smaller scale and tends to focus on the use of natural materials that are available locally. Other commonly used terms include sustainable design and green architecture; however, while good design is essential to green building, the actual operation, maintenance, and ultimate disposal or deconstruction of the building also have very significant effects on buildings' overall environmental impact.

The related concepts of sustainable development and sustainability are integral to green building. Effective green building can lead to 1) *reduced operating costs* by increasing productivity and using less energy and water, 2) *improved public and occupant health* due to improved indoor air quality, and 3) *reduced environmental impacts* by, for example, lessening storm water runoff and the heat island effect. Practitioners of green building often seek to achieve not only ecological but aesthetic harmony between a structure and its surrounding natural and built environment. The sustainable buildings are also environmentally friendly in the fact that they are built out of materials that are good for the environment.The appearance and style of sustainable homes and buildings can be nearly indistinguishable from their less sustainable counterparts.

Green buildings are scored by rating systems, such as the Leadership in Energy and Environmental Design (LEED) rating system developed by the U.S. Green Building Council, Green Globes from GBI and other locally developed rating systems.

The Environmental Impact of Buildings

Buildings have a profound effect on the environment, which is why green building practices are so important to reduce and perhaps one day eliminate those impacts.

In the United States alone, buildings account for:

- 39% of total energy use
- 12% of total water consumption
- 68% of total electricity consumption
- 38% of total carbon dioxide emissions

However, the environmental impact of buildings is often underestimated, while the perceived costs of building green are overestimated. A recent survey by the World Business Council for Sustainable Development finds that green costs are overestimated by 300%, as key players in real estate and construction estimate the additional cost at 17% above conventional construction, more than triple the true average cost difference of about 5%.

Green Building Practices

Green building brings together a vast array of practices and techniques to reduce and ultimately eliminate the impacts of buildings on the environment. On the aesthetic side of green architecture or sustainable design is the philosophy of designing a building that is in harmony with the natural features and resources surrounding the site. There are several key steps in designing sustainable buildings: specify 'green' building materials from local sources, reduce loads, optimize systems, and generate on-site renewable energy.

Building materials typically considered to be 'green' include rapidly renewable plant materials like bamboo and straw, lumber from forests certified to be sustainably managed, stone, recycled metal, and other products that are non-toxic, reusable, renewable, and/or recyclable. Building materials should be extracted and manufactured locally to the building site to minimize the energy embedded in their transportation.

Low-impact building materials are used wherever feasible: for example, insulation may be made from low VOC (volatile organic compound)-emitting materials such as recycled denim, rather than the insulation materials that may contain carcinogenic or toxic materials such as formaldehyde. To discourage insect damage, these alternate insulation materials may be treated with boric acid. Organic or milk-based paints may be used.

Architectural salvage and reclaimed materials are used when appropriate as well. When older buildings are demolished, frequently any good wood is reclaimed, renewed, and sold as flooring. Many other parts are reused as well, such as doors, windows, mantels, and hardware, thus reducing the consumption of new goods. When new materials are employed, green designers look for materials that are rapidly replenished, such as bamboo, which can be harvested for commercial use after only 6 years of growth, or cork oak, in which only the outer bark is removed for use, thus preserving the tree. When possible, building materials may be gleaned from the site itself; for example, if a new structure is being constructed in a wooded area, wood from the trees which were cut to make room for the building would be re-used as part of the building itself.

To minimize the energy loads within and on the structure, it is critical to orient the building to take advantage of cooling breezes and sunlight. Daylighting with ample windows will eliminate the need to turn on electric lights during the day (and provide great views outside too). Passive Solar can warm a building in the winter - but care needs to be taken to provide shade in the summer time to prevent overheating. Prevailing breezes and convection currents can passively cool the building in the summer. Thermal mass stores heat gained during the day and releases it at night minimizing the swings in temperature. Thermal mass can both heat the building in winter and cool it during the summer. Insulation is the final step to optimizing the structure. Well-insulated windows, doors, and walls help reduce energy loss, thereby reducing energy usage. These design features don't cost much money to construct and significantly reduce the energy needed to make the building comfortable.

Optimizing the heating and cooling systems through installing energy efficient machinery, commissioning, and heat recovery is the next step. Compared to optimizing the passive heating and cooling features through design, the gains made by engineering are relatively expensive and can add significantly to the projects cost. However, thoughtful integrated design can reduce costs — for example, once a building has been designed to be more energy-efficient, it may be possible to downsize heating, ventilation and air-conditioning (HVAC) equipment, leading to substantial savings. To further address energy loss hot water heat recycling is used to reduce energy usage for domestic water heating. Ground source heat pumps are more energy efficient then other forms of heating and cooling.

Finally, onsite generation of renewable energy through solar power, wind power, hydro power, or biomass can significantly reduce the environmental impact of the building. Power generation is the most expensive feature to add to a building.

Good green architecture also reduces waste, of energy, water and materials. During the construction phase, one goal should be to reduce the amount of material going to landfills. Well-designed buildings also help reduce the amount of waste generated by the occupants as well, by providing onsite solutions such as compost bins to reduce matter going to landfills.

To reduce the impact on wells or water treatment plants, several options exist. "Greywater", wastewater from sources such as dishwashing or washing machines, can be used for subsurface irrigation, or if treated, for non-potable purposes, e.g., to flush toilets and wash cars. Rainwater collectors are used for similar purposes.

Green building often emphasizes taking advantage of renewable resources, e.g., using sunlight through passive solar, active solar, and photovoltaic techniques and using plants and trees through green roofs, rain gardens, and for reduction of rainwater run-off.[4] Many other techniques, such as using packed gravel for parking lots instead of concrete or asphalt to enhance replenishment of ground water, are used as well.

Green Building Worldwide

Standards and Ratings

Many countries have developed their own standards of energy efficiency for buildings.

- Code for Sustainable Homes, United Kingdom
- BREEAM, United Kingdom
- EnerGuide for Houses, Canada (energy retrofits & up-grades)
- EnerGuide for New Houses, Canada (new construction)
- Gold & Silver Energy Standards, United Kingdom
- Green Building Council of Australia's Green Star
- Haute Qualité Environnementale, France
- House Energy Rating, Australia
- Leadership in Energy and Environmental Design (LEED), USA, Canada & INDIA
- Green Globes, USA, Canada and United Kingdom
- Minergie, Switzerland

- National Association of Home Builders Green Building Guidelines, USA
- New Zealand Green Building Council Green Star
- *Passivhaus*, Germany, Austria, United Kingdom
- *EEWH*, Taiwan

Australia

There is a system in place in Australia called First Rate designed to increase energy efficiency of residential buildings. The Green Building Council of Australia (GBCA) has developed a green building standard known as Green Star.

In Adelaide, South Australia, there are at least two different projects that incorporate the principles of Green building. The Eco-City development is located in Adelaide's city centre and the Aldinga Arts Eco Village is located in Aldinga. Guidelines for building developments in each project are outlined in the bylaws. The bylaws include grey water reuse, reuse of stormwater, capture of rainwater, use of solar panels for electricity and hotwater, solar passive building design and community gardens and landscaping.

Melbourne has a rapidly growing environmental consciousness, many government subsidies and rebates are available for water tanks, water efficient products (such as shower heads) and solar hot water systems. The city is home to many examples of green buildings and sustainable development such as the CERES Environmental Park. Another one is EcoLinc in Bacchus Marsh. Two of the most prominent examples of green commercial buildings in Australia are located in Melbourne - 60L and Council House 2 (also known as CH2).

The most recent building to receive the 6 Green Star award was in Canberra, where Australian Ethical Investment Ltd refurbished an existing office space in Trevor Pearcey House. The total cost of the renovation was $1.7 million, and produced an estimated 75% reductions in carbon dioxide emissions, 75% reduction in water usage, and used over 80% recycled materials. The architects were Collard Clarke Jackson Canberra, architectual work done by Kevin Miller, interior design by Katy Mutton.

Canada

Canada has implemented "R-2000" guidelines for new buildings built after the year 2000. Incentives are offered to builders to meet the R-2000 standard in an effort to increase energy efficiency and promote sustainability.

A progression of the R-2000 home program is the EnerGuide for New Houses service. This service is available across Canada and is designed to allow home builders and home buyers to build homes that use significantly less energy than the average homes being built. Some Canadian provinces are considering mandatory use of the service for all new homes.

In December 2002, Canada formed the Canada Green Building Council and in July 2003 obtained an exclusive licence from the US Green Building Council to adapt the LEED rating system to Canadian circumstances.

- Beamish-Munro Hall at Queen's University features sustainable construction methods such as high fly-ash concrete, triple-glazed windows, dimmable fluorescent lights and a grid-tied photovoltaic array.
- Gene H. Kruger Pavilion at Laval University uses largely non polluting, non toxic, recycled and renewable materials as well as advanced bioclimatic concepts that reduce energy consumption by 25% compared with a concrete building of the same dimensions. The structure of the building is made entirely out of wood products, thus further reducing the environmental impact of the building.

Germany

German developments that employ green building techniques include:

- The Solarsiedlung (Solar Village) in Freiburg, Germany, which features energy-plus houses.
- The Vauban development, also in Freiburg.
- Houses designed by Baufritz, incorporating passive solar design, heavily insulated walls, triple-glaze doors and windows, non-toxic paints and finishes, summer shading, heat recovery ventilation, and greywater treatment systems.
- The new Reichstag building in Berlin, which produces its own energy.

India

The Confederation of Indian Industry (CII) plays an active role in promoting sustainability in the Indian construction sector. The CII is the central pillar of the Indian Green Building Council or IGBC. The IGBC has licesensed the LEED Green Building Standard from the U.S. Green Building Council and currently is responsible for certifying LEED-New Construction and LEED-Core and Shell buildings in India. All other projects are certified through the U.S. Green Building Council. There are many energy efficient buildings in India, situated in a variety of climatic zones.

Israel

Israel has recently implemented a voluntary standard for "Buildings with Reduced

Environmental Impact" 5281, this standard is based on a point rating system (55= certified 75=excellence) and together with complementary standards 5282-1 5282-2 for energy analysis and 1738 for sustainable products provides a system for evaluating environmental sustainability of buildings. United States Green Building Council LEED rating system has been implemented on several building in Israel including the recent Intel Development Center in Haifa and there is strong industry drive to introduce an Israeli version of LEED in the very near future.

Malaysia

The Standards and Industrial Research Institute of Malaysia (SIRIM) promotes green building techniques. Malaysian architect Ken Yeang is a prominent voice in the area of ecological design.

New Zealand

The New Zealand Green Building Council has been in formation since July 2005. An establishment board was formed later in 2005 and with formal organisational status granted on 1st February 2006. That month Jane Henley was appointed as the CEO and activity to gain membership of the World GBC began. In July 2006 the first full board was appointed with 12 members reflecting wide industry involvement. The several major milestones were achieved in 2006/2007; becoming a member of the World GBC, the launch of the Green Star NZ - Office Design Tool, and welcoming our member companies.

United Kingdom

The Association for Environment Conscious Building (AECB) has promoted sustainable building in the UK since 1989.

The UK Building Regulations set requirements for insulation levels and other aspects of sustainability in building construction.

United States

The United States Green Building Council (USGBC) has developed The Leadership in Energy and Environmental Design (LEED) Green Building Rating System™, which is the nationally accepted benchmark for the design, construction, and operation of high performance green buildings. LEED gives building owners and operators the tools they need to have an immediate and measurable impact on their buildings' performance. LEED promotes a whole-building approach to sustainability by recognizing performance in five key areas of human and environmental health: sustainable site development, water savings, energy efficiency, materials selection, and indoor environmental quality. They have developed specific versions of the

LEED rating system to assist specific building types in achieving certification. Some of the commercially available systems are:

- LEED-NC: New Construction and Major Renovations (the most commonly applied-for LEED certification)
- LEED-CI: Commercial Interiors
- LEED-CS: Core/Shell)
- LEED-EB: Existing Buildings
- LEED-Homes

Other versions that will soon be released for public consumption are:

- LEED-ND: Neighborhood Developments
- LEED for Schools
- LEED for Healthcare
- LEED for Labs
- LEED for Retail

The Green Building Initiative is a non-profit network of building industry leaders committed to bringing green to mainstream residential and commercial construction. The GBI believes in building approaches that are environmentally progressive, but also practical and affordable for builders to implement. The GBI has developed an easy to use, inexpensive and web-based rating tool called Green Globes, which is being upgraded in accordance with ANSI procedures.

The United States Environmental Protection Agency's EnergyStar program rates commercial buildings for energy efficiency and provides EnergyStar qualifications for new homes that meet their standards for energy efficient building design.

ShoreBank, an American community development bank, has pioneered lending to developers for Green Building projects and energy audits of existing structures.

In 2005, Washington became the first state in the U.S. to enact green building legislation. According to the law, all major public agency facilities with a floor area exceeding 5,000 square feet (465 m²), including state funded school buildings, are required to meet or exceed LEED standards in construction or renovation. The projected benefits from this law are:

- 20% annual savings in energy costs
- 20% reduction in water costs

- 38% reduction in waste water production
- 22% reduction in construction waste

In 2006, Charlottesville, VA became one of the first small towns in the US to enact green building legislation. This presents a significant shift in construction and architecture as LEED regulations have formerly been focused on commercial construction. If US homeowner interest grows in "green" residential construction, the companies involved in the production and manufacturing of LEED building materials will become likely candidates for tomorrow's round of private equity and IPO investing.

GREEN COMPUTING

Green computing is the study and practice of using computing resources efficiently. Typically, technological systems or computing products that incorporate green computing principles take into account the so-called triple bottom line of economic viability, social responsibility, and environmental impact. This differs somewhat from traditional or standard business practices that focus mainly on the economic viability of a computing solution. These focuses are similar to those of green chemistry; reduction of the use of hazardous materials such as lead at the manufacturing stage, maximized energy efficiency during the product's term of use, and recyclability or biodegradability of both a defunct product and of any factory waste.

A typical green computing solution attempts to address some or all of these factors by implementing environmentally friendly products in an efficient system. For example, an IT manager might purchase Electronic Products Environmental Assessment Tool (EPEAT)-approved hardware combined with a thin client solution. As compared to a traditional desktop PC configuration, such a configuration would probably reduce IT maintenance-related activities, extend the useful life of the hardware, and allow for responsible recycling of the equipment past its useful life.

Origins

In 1992, the U.S. Environmental Protection Agency launched Energy Star, a voluntary labeling program, designed to promote and recognize energy-efficiency in monitors, climate control equipment, and other technologies. This resulted in, among other things, the widespread adoption of sleep mode among consumer electronics. The term "green computing" was probably coined shortly after the Energy Star program began, and generally referred to power consumption-related issues. There are several USENET posts dating back to 1992 which use the term in this manner. Also in 1992, the Swedish organization TCO Development launched the TCO Certification program to promote low magnetic and electrical emissions from CRT-based computer displays. The program was later expanded to include criteria on energy consumption, ergonomics, and the use of hazardous materials in construction.

Fueled by recent trends towards sustainability, the modern use of the term refers to systematic approaches to using computing technology efficiently. These include items such as addressing electronic waste, regulatory compliance, telecommuting policies, virtualization of server resources, cost accounting of energy use, thin client solutions, and many others.

Rationale

There are different philosophies related the implementation of green computing solutions. Virtually all proponents believe that businesses need to reevaluate some of their technological solutions and/or policies.

- Tactical incrementalists tend to preserve existing IT infrastructure and policies but will incorporate some environmental and social factors into future decisions. An example of this is to encourage power management policies for computing equipment. Changes of this type are generally easy to implement, are not part of a unified plan, and require little political effort. Similarly to a standard business model, fiscal viability remains the primary concern, with little or no value being attributed to the environmental or social factors.
- Strategic leaders recognize that environmental and social factors are contributing to a trend in disruptive technology. As such, there is considerable value to altering the existing infrastructure and policies. While the primary rationale is still cost efficiency, other factors such as branding, marketing, or hiring may be referenced as well. A baseline study will probably be conducted, and more complex, pronounced initiatives will be implemented. For example, the IT department may be held accountable for the cost of all electricity from computer related equipment, or desktop computing will be replaced with thin client computing.
- Deep green technologists believe that computing has reached a point of diminishing returns and needs to be substantially reevaluated. Many are of the opinion that the dawn of the 21st century resulted in a rush to technology, which has resulted in an ongoing hubris for computing initiatives that are fiscally unfounded and environmentally destructive. This hysteria shows itself in many forms; ROI, for example, is rarely calculated for IT projects across the industry, and the growing stocks of electronic waste throughout the world is reaching crisis levels. Unlike the other categories, social and environmental factors are construed as being on equal footing with fiscal responsibility. For example, a deep green technologist might implement a

carbon offset policy to neutralize the cost of electricity for computing, or use solar powered web hosting services.

Regulations and Industry Initiatives

Government

Many governmental agencies and have implemented standards and regulations that encourage green computing. The U.S. Environmental Protection Agency's Energy Star program, which was launched in 1992, was revised in October 2006 to include stricter efficiency requirements for computer equipment, and a tiered ranking system for approved products. In Europe, the Swedish Confederation of Professional Employees certifies personal computers, monitors, and other office equipment that meets the ergonomics, energy usage, emissions, and hazardous substances requirements of the TCO Certification program. The European Union's directives 2002/95/EC (RoHS), on the reduction of hazardous substances, and 2002/96/EC (WEEE) on waste electrical and electronic equipment required the substitution of heavy metals and flame retardants like PBBs and PBDEs in all electronic equipment put on the market starting on 2006-07-01. The directives placed responsibility on manufacturers for the gathering and recycling of old equipment.

In 2003, the California State Senate enacted the Electronic Waste Recycling Act, which establishes a state-wide recycling program for obsolete computer and consumer electronics equipment. To pay for the program, the Act imposes a fee for each unit sold at retail, based on the size of the display. The Act also establishes restrictions on hazardous substances, equal to the European Union's RoHS Directive.

Industry

The Green Electronics Council offers the Electronic Products Environmental Assesment Tool (EPEAT) to assist in the purchase of "green" computing systems. The Council evaluates computing equipment on 28 criteria that measure a product's efficiency and sustainability attributes. On 2007-01-24, President George W. Bush issued Executive Order 13423, which requires all United States Federal agencies to use EPEAT when purchasing computer systems.

In May 2007, IBM launched Project Big Green, a 1 billion-USD-per-year effort to design and promote energy efficiency in corporate data centers.

Approaches to Green Computing

Several methods are available for minimizing the environmental impact of a computing system.

Power Generation

All computers require electrical power to operate, regardless of the system's efficiency. One of the goals of green computing is to use power generated from sources that are more environmentally friendly than coal-fired power stations. Some companies provide power from low impact sources, like windmills and hydroelectric dams. Other sources include locally-installed photovoltaic panels, which generate electrical energy from the Sun, or power produced by nuclear power plants.

A laptop can be even powered by human power which brings additional exercise, like in the Twibright Exciter project.

Virtualization

Computer virtualization is the process of running two or more logical computer systems on one set of physical hardware. The concept originated with the mainframe operating systems of the 1960s, but was commercialized for x86-compatible computers only in the 1990s. With virtualization, a system administrator could combine several physical systems into virtual machines on one single, powerful system, thereby unplugging the original hardware and reducing power and cooling consumption. Several commercial companies and open-source projects now offer software packages to enable a transition to virtual computing. Intel Corporation and AMD have also built proprietary virtualization enhancements to the x86 instruction set into each of their CPU product lines, in order to facilitate virtualized computing.

Power Management

An open industry standard called Advanced Configuration and Power Interface (ACPI) provides a standard programming interface that allows an operating system to directly control the power saving aspects of the hardware. This allows the system to automatically turn off components such as monitors and hard drives after set periods of inactivity. In addition, a system may hibernate, in which it turns off nearly all components, including the CPU and the system RAM, greatly reducing the system's electricity usage. To resume from this state, some components, such as the keyboard, network interface card, and USB ports may remain powered, to receive input from the user. ACPI itself is a successor to an earlier Intel-Microsoft standard called Advanced Power Management, which allows a computer's BIOS to control power management functions.

In the absence of ACPI or APM support, some external components, such as computer displays, printers, scanners, speakers, and hard drives may be turned off manually when not in use. In this state, though the external periphals may be off, the main system continues to consume electricity. To minimize the impact, the system could run file sharing software or volunteer computing software, donating its resources to a long-term project.

Some software programs allow the user to manually adjust the voltages supplied to the CPU, essentially reducing the amount of electricity used by the CPU while it's on and powered. Since many CPUs have "safety-nets" on either side of the spectrum (+/- the voltage parameters of a given CPU), one is able to reduce the amount of volts the processor uses, hence reducing both the amount of heat produced and the amount of electricity consumed. Some CPUs from Intel Corporation and AMD, particularly those intended for use in laptops, have technology to automatically adjust the processor voltages depending on the workload. This technology is called "SpeedStep" with intel processors, "PowerNow!"/"Cool'n'Quiet" with AMD chips, LongHaul with VIA CPUs, and LongRun with Transmeta processors.

In 2007, Intel Corporation released a utility called PowerTOP, which measures and reports on a PC's power consumption. This utility is available only for PCs running a Linux operating system.

Newer Hardware

As of 2007, at least four personal computer vendors (Everex, Linutop, Systemax, and Zonbu) ship dedicated low-power thin client PCs. These systems are desinged primarily for the purpose of reducing power consumption, by providing minimal hardware peripherals and low-voltage components. These design decisions limit the computational performance of such machines, which makes them impractical for applications that require a lot of processing power, such as computer gaming and video production. Dedicated low-power PCs generally are physically much smaller than traditional desktop PCs, and as such can be mounted under desks or on walls, creating more usable desk space. The Geode processor from AMD (originally from National Semiconductor), is one example of a CPU that strives for minimal power consumption while maintaining full compatibility with the x86 instruction set.

Solid state drives store persistent data in either flash memory or DRAM, which do not need any mechanical parts, unlike traditional platter-based hard disk drives. The spindle motor and the data platters are primarily responsible for a hard disk drive's power consumption.

LCD monitors typically use a cold-cathode fluorescent bulb to provide light for the display. Newer displays use an array of light-emitting diodes (LEDs) in place of the fluorescent bulb, which reduces the amount of electricity used by the display. This also reduces the amount of hazardous substances in the system, as all fluorescent bulbs contain trace amounts of mercury.

Desktop computer power supplies (PSUs) are generally about 70–75% efficient;[12] to produce 75W of DC output they require 100W of AC input and dissipate the remaining 25W in heat. An industry initiative called 80 PLUS certifies PSUs that are at least 80% efficient. Typically these models are drop-in replacements for older, non-certified PSUs of the same form factor. As of 2007-07-20, all new Energy Star 4.0-certified desktop PSUs must be at least 80% efficient.

Materials Recycling

Obsolete, but still functional computer systems can be donated to various charities and non-profit organizations. Many charities have minimum system requirements for acceptable computer systems. Broken systems, or those too old to be useful to charities, can be recycled through some retail outlets and municipal or private recycling centers, often at a fixed cost per item.

Recycling old computers has the benefit of keeping defunct systems - and the harmful materials contained within, such as lead, mercury, hexavalent chromium, among others - out of the regular waste stream and thus, out of landfills. However, some observers note that some old computers, gathered through recycling drives, end up shipped to developing countries, where environmental standards are less strict than in North America and Europe. Columnist Sascha Segan wrote in PC Magazine that there is little accountability for where computers go after being collected for recycling from the consumer. In her book *Garbage Land*, author Elizabeth Royte quoted the Silicon Valley Toxics Coaltion, which stated that 80% of the post-consumer e-waste collected for recycling is shipped to countries such as China, India, and Pakistan, where the electronic devices are manually broken apart and sifted to collect precious metals, such as copper and gold.

HYDROGEN TECHNOLOGIES

Hydrogen technologies are technologies for creating electricity from hydrogen. They are related to the goal of a future hydrogen economy. The application of hydrogen technologies have the potential to play role in preventing climate change and are applicable in many uses.

The following list gives links to articles related to hydrogen technologies:

Hydrogen Production

- Artificial photosynthesis
- Biological hydrogen production (Algae)
- Electrolysis

- Electrolysis electrochemical cell
- Steam reforming
- Sulfur-iodine cycle
- Water splitting

Fuel Cells

- Alkaline fuel cell (AFC)
- Direct borohydride fuel cell (DBFC)
- Direct-ethanol fuel cell (DEFC)
- Direct methanol fuel cell (DMFC)
- Electro-galvanic fuel cell (EGFC)
- Flow Battery (RFC)
- Formic acid fuel cell (FAFC)
- Metal Hydride fuel cell (MHFC)
- Microbial Fuel Cell (MFC)
- Molten-carbonate fuel cell (MCFC)
- Phosphoric-acid fuel cell (PAFC)
- Photoelectrochemical cell (PEC)
- Proton exchange membrane fuel cell (PEMFC)
- Protonic Ceramic Fuel Cell (PCFC)
- Reversible fuel cell (RFC)
- Solid-oxide fuel cell (SOFC)

Hydrogen Infrastructure

- Hydrogen compressor
- Hydrogen highway
- Hydrogen leak testing
- Hydrogen microsensor
- Hydrogen analyzer
- Hydrogen piping
- Home Energy Station
- Zero Regio

Hydrogen Storage

- Liquid hydrogen
- Hydrogen station

Hydrogen Vehicles

Historic Hydrogen Filled Airships

- Hindenburg (airship)

Hydrogen Powered Cars

- BMW H2R
- BMW Hydrogen 7
- Chrysler Natrium
- Fiat Panda Hydrogen
- Fuel Cell Bus Club
- General Motors Hy-wire
- General Motors Sequel
- Honda FCX
- Mercedes-Benz F-Cell
- Morgan LIFEcar
- Peugeot Quark

Hydrogen Powered Planes

- Hyfish
- Smartfish
- Tupolev Tu-155-hydrogen-powered version of Tu-154

Related Technologies

Hydrogen is not in and of itself an energy source, because it is not naturally occurring as a fuel. The potential environmental impact depends primarily on the methods used to generate the hydrogen fuel.

Environmental

- Anaerobic digestion
- Syngas

Nuclear

- Generation IV reactor
- Hydrogen bomb

Organic Chemistry:

- Dehydrogenation
- Hydrogenation
- Hydrogenolysis

Miscellaneous

- Atomic hydrogen welding
- Oxyhydrogen flame
- Standard hydrogen electrode
- Palladium-Hydrogen electrode
- Cathodic protection
- Hofmann voltameter
- Joule-Thomson effect
- Hydrogen ion
- Bussard ramjet
- Döbereiner's lamp
- Gas-absorption refrigerator
- Diaphragm compressor
- Metal Diaphragm
- Thermal hydrogen compressor
- Electrochemical compression
- Hydrogen damage
- Hydrogen embrittlement
- Hydrogen grooving

REFERENCES

Evans, Ianto, Michael G. Smith, and Linda Smiley. 2003. *The Hand-Sculpted House: A Practical and Philosophical Guide to Building a Cob Cottage.* Chelsea Green Publishing.

Kennedy, J., Smith, M. and Wanek, C. eds. 2002. *The Art of Natural Building: Design, Construction, Resources.* Gabriola Island, BC. New Society Publishers.

Lynne Elizabeth, L., and Adams, C., eds. 2000. *Alternative Construction, Contemporary Natural Building Methods* pub. by John Wiley & Sons, inc. NY, NY

Woolley T. 2006. *Natural Building: A Guide to Materials and Techniques.* Crowood Press.

6

Recycling, Waste and Wastewater Treatment & Management

RECYCLING

Recycling is the reprocessing of materials into new products. Recycling generally prevents the waste of potentially useful materials, reduces the consumption of raw materials and reduces energy usage, and hence greenhouse gas emissions, compared to virgin production. Recycling is a key concept of modern waste management and is the third component of the waste hierarchy.

Recyclable materials, also called “recyclables”, may originate from a wide range of sources including the home and industry. They include glass, paper, aluminium, asphalt, iron, textiles and plastics. Biodegradable waste, such as food waste or garden waste, is also recyclable with the assistance of micro-organisms through composting or anaerobic digestion.

Recyclates are sorted and separated into material types. Contamination of the recylates with other materials must be prevented to increase the recyclates’ value and facilitate easier reprocessing for the ultimate recycling facility. This sorting can be performed either by the producer of the waste or within semi- or fully-automated materials recovery facilities.

There are two common household methods of recycling. In curbside collection (UK: kerbside collection), consumers leave presorted recyclable materials in front of their property to be collected by a recycling vehicle. With a “bring” or carry-in system, the householder takes the materials to collection points, such as transfer stations or civic amenity sites.

The term *recycling* does not generally include reuse, in which existing items are used for a new purpose.

History

Recycling has been a common practice throughout human history. In pre-industrial times, scrap made of bronze and other precious metals was collected in Europe and melted down for perpetual reuse, and in Britain dust and ash from wood and coal fires was downcycled as a base material in brick making. The main driver for these types of recycling was the economic advantage of obtaining recycled feedstock instead of acquiring virgin material, as well as a lack of public waste removal in ever more-populated sites.

Paper recycling began in Britain in 1921, when the British Waste Paper Association (now Confederation of Paper Industries) was established to encourage trade in waste paper recycling.

Resource shortages caused by the world wars, and other such world-changing occurrences greatly encouraged recycling. Massive government promotion campaigns were carried out in World War II in every country involved in the war, urging citizens to donate metals and conserve fiber, as a matter of significant patriotic importance. Resource conservation programs established during the war were continued in some countries without an abundance of natural resources, such as Japan, after the war ended.

The next big investment in recycling occurred in the 1970s, due to rising energy costs. Recycling aluminum uses only 5% of the energy required by virgin production; glass, paper and metals have less dramatic but very significant energy savings when recycled feedstock is used. The passage of the Clean Water Act of 1977 in the USA created strong demand for bleached paper (office paper whose fibre has already been bleached white increased in value as water effluent became more expensive).

In 1973, the city of Berkeley, California began one of the first curbside collection programs with monthly pick ups of newspapers from residences. Since then several countries have started and expanded various doorstep collection schemes. Around this time, Woodbury, New Jersey was also a forerunner of the recycling industry in the United States, being the first in the state to mandate it.

In 1987, the Mobro 4000 barge hauled garbage from New York to North Carolina, where it had been denied. It was then sent to Belize, where it was denied as well. Finally, the barge returned to New York and the garbage was incinerated. The incident led to heated discussions in the media about waste disposal and recycling. The incident is often referred to as igniting the recycling hysteria of the 1990s.

One event that initiated recycling efforts occurred in 1989 when Berkeley banned the use of polystyrene packaging for keeping McDonald's hamburgers warm. One effect of this ban was to raise the ire of management at Dow Chemical, the world's largest manufacturer of polystyrene, which led to the first major effort to show that plastics can be recycled. By 1999, there were 1,677 companies in the USA alone involved in the post-consumer plastics recycling business.

In Germany the introduction of a Packaging Ordinance under the Waste Act led to the introduction of the Green Dot program. Originally launched by Duales System Deutschland GmbH (DSD) in 1990, the scheme has since been rolled out to 23 other European countries. The Green Dot scheme is captured under the European "Packaging and Packaging Waste Directive - 94/62/EC" and is binding for all companies if their products use packaging.

Benefits

Recycling is beneficial in two ways: it reduces the inputs (energy and raw materials) to a production system and reduces the amount of waste produced for disposal.

A study conducted by the Technical University of Denmark found that in 80% of cases, recycling is the most efficient method to dispose of household waste.

Some materials like aluminum can be recycled indefinitely as there is no change to the materials. Other recycled materials like paper require a percentage of raw materials (wood fibers) to be added to compensate for the degradation of existing fibers. Recycling aluminium saves 95% of the energy cost of processing new aluminium because the melting temperature is reduced from 900 °C to 600 °C. It is by far the most efficient material to recycle. Recycling plastic saves 70% of the energy used in creating new plastic, and paper recycling saves 40% of the energy required to make a new product.

The resources being processed are purer, thus less energy is needed to process them. Additionally less energy is needed to transport them from the place of extraction (e.g. bauxite/aluminium ore mines in Brazil or coniferous forests in Scandinavia as compared to domestic areas).

This reduces the environmental, social, and usually the economic costs of manufacturing.

For example, bauxite mines in Brazil displace indigenous people, create noise pollution from blasting, machinery and transport, and create air pollution in the form of particulates (dust). The habitat loss and visual destruction is also negative both to the aesthetic qualities of the areas and the local environment.

The most commonly used methods for waste disposal (landfill, pyrolysis, incineration) may be environmentally damaging and unsustainable. Therefore any way to reduce the volume of waste being disposed in this fashion may be beneficial. The maximum environmental benefit is gained by waste minimization (reducing the amount of waste produced), and reusing items in their current form such as refilling bottles.

In the past recycling has been a fragmented industry with recycling companies only accepting specific materials for recycling, or only common materials such as plastic or metals. More recent approaches have resulted in recycling companies that accept a wide variety of materials or emerging market materials such as ceiling tiles, carpet, stained wood etc.. An example of a company that accepts a wide variety of materials, including emerging recycling markets is the Green Recycling Network.

Drawbacks and Criticism

All recycling techniques consume energy for transportation and processing and some also use considerable amounts of water, although recycling processes seldom amount to the level of resource use associated with raw materials processing.

There may also be drawbacks with the collection methods associated with recycling. Increasing collections of separated wastes adds to vehicle movements and the production of carbon dioxide. This may be negated however by centralized facilities such as some advanced material recovery facilities and mechanical biological treatment systems for the separation of mixed wastes. It has been calculated that collecting waste and disposing it in a landfill is about $60 a ton opposed to separate collecting and taking it to be recycled costs $150 a ton.

Recycled materials also sometimes cost more financially than their non-recycled versions. This is not universal to every recycled product, but it does occur.

Negative consequences from mercury recycling have been cited by *The Wall Street Journal*. The article traces mercury recovered from American recycling programs into sales of mercury for alluvial mining activities in Brazil. During the autumn of 2006, the European Union banned the export of liquid mercury, and a life-cycle analysis prior to institution of recycling programs may reduce the risk of unintended environmental consequences.

For some materials, recycling is more expensive than landfill disposal unless externalities are considered.[citation needed] For these products, there is an opportunity cost to recycling. Some argue that the financial costs of recycling some materials outweigh the environmental benefits. Some skeptics also argue that the environmental benefits of recycling do not compensate for the extra effort it may require.

According to an article in The New York Times by John Tierney, government mandated recycling wastes more resources than it saves. Some highlights from the article:

- In cases where recycling truly does save resources, such as with large scraps of aluminum, this will be reflected in market prices, and voluntary recycling will take place. Thus, there is no need for the government to mandate it.
- Each year the United States fills up less than 10 square miles of landfill space. Once full, much of that land gets turned into parks.
- Tree farmers plant more trees than they cut down.
- Government mandated recycling is more expensive than putting the garbage into landfills, which means that this recycling uses up more resources than it saves.
- Some small towns with landfills are happy to import garbage from other cities and states because it provides jobs and tax revenue.
- Today's modern landfills are much cleaner and safer, and much less likely to leak and pollute, than the landfills of the past.

Recycling Techniques

Many different materials can be recycled but each type requires a different technique.

Aggregates & Concrete

Concrete aggregate collected from demolition sites is put through a crushing machine, often along with asphalt, bricks, dirt, and rocks. Smaller pieces of concrete are used as gravel for new construction projects. Crushed recycled concrete can also be used as the dry aggregate for brand new concrete if it is free of contaminants. This reduces the need for other rocks to be dug up, which in turn saves trees and habitats.

Batteries

The large variation in size and type of batteries makes their recycling extremely difficult: they must first be sorted into similar kinds and each kind requires an individual recycling process. Additionally, older batteries contain mercury and cadmium, harmful materials which must be handled with care.

Lead-acid batteries, like those used in automobiles, are relatively easy to recycle and many new lead-acid batteries contain a high percentage of recycled material.

Biodegradable Waste

Biodegradable waste can be recycled into useful material by biological decomposition. There are two mechanisms by which this can occur. The most common mechanism of recycling of household organic waste is home composting or municipal curbside collection of green wastes sent to large scale composting plants.

Alternatively organic waste can be converted into biogas and soil improver using anaerobic digestion. Here organic wastes are broken down by anaerobic microorganisms in biogas plants. The biogas can be converted into renewable electricity or burnt for environmentally friendly heating. Advanced technologies such as mechanical biological treatment are able to sort the recyclable elements of the waste out before biological treatment by either composting, anaerobic digestion or biodryin

Sulabh International, an Indian NGO designed a cheap toilet system that recycles human waste into biogas and fertiliser

Electronics Disassembly and Reclamation

The direct disposal of electrical equipment — such as old computers and mobile phones — is banned in many areas due to the toxic contents of certain components. The recycling process works by mechanically separating the metals, plastics and circuit boards contained in the appliance. When this is done on a large scale at an electronic waste recycling plant, component recovery can be achieved in a cost-effective manner.

Electronic devices, including audio-visual components (televisions, VCRs, stereo equipment), mobile phones and other hand-held devices, and computer components, contain valuable elements and substances suitable for reclamation, including lead, copper, and gold. They also contain a plethora of toxic substances, such as dioxins, PCBs, cadmium, chromium, radioactive isotopes, and mercury. Additionally, the processing required to reclaim the precious substances (including incineration and acid treatments) release, generate and synthesize further toxic by-products.

In the United States, an estimated 70% of heavy metals in landfills come from discarded electronics. Some regional governments are attempting to curtail the accumulation of electronics in landfills by passing laws obligating manufacturers and consumers to recycle these devices, but because in many cases safe dismantlement of these devices in accordance with first world safety standards is unprofitable, historically much of the electronic waste has been shipped to countries with lower or less rigorously-enforced safety protocols. Places like Guiyu, China dismantle tons of electronics every year, profiting from the sale of precious metals, but at the cost of the local environment and the health of its residents.

Mining to produce the same metals, to meet demand for finished products in the west, also occurs in the same countries, and the United Nations Conference on Trade and Development (UNCTAD) has recommended that restrictions against recycling exports be balanced against the environmental costs of recovering those materials from mining. Hard rock mining in the USA produces 45% of all toxics produced by all USA industries (2001 US EPA Toxics Release Inventory).

Printer Ink Cartridges & Toners

Printer ink cartridges can be recycled. They are sorted into different brands and models which are then resold back to the companies that created these cartridges. The companies then refill the ink reservoir which can be sold back to consumers. Toner cartridges are recycled the same way as ink cartridges, using toner instead of ink. This method of recycling is highly efficient as there is no energy spent on melting and recreating the recycled object itself.

Ferrous Metals

Iron and steel are the world's most recycled materials, and among the easiest materials to recycle, as they can be separated magnetically from the waste stream. Recycling is via a steelworks: scrap is either remelted in an Electric Arc Furnace (90-100% scrap), or used as part of the charge in a Basic Oxygen Furnace (around 25% scrap). Any grade of steel can be recycled to top quality new metal, with no 'downgrading' from prime to lower quality materials as steel is recycled repeatedly. 42% of crude steel produced is recycled material.

Non-Ferrous Metals

Aluminium is shredded and ground into small pieces or crushed into bales. These pieces or bales are melted in an aluminium smelter to produce molten aluminium. By this stage the recycled aluminium is indistinguishable from virgin aluminium and further processing is identical for both.

Due to the high melting point of aluminium ore, large amounts of energy are required to extract aluminium from ore, making the environmental benefits of recycling aluminium enormous. Recycling aluminium only results in approximately 5% of the CO_2 that would be released during the production of raw aluminium. The percentage is even smaller when considering the complete cycle of mining and transporting the aluminium. Also, as open-cut mining is most often used for obtaining aluminium ore, mining destroys large sections of natural land.

An aluminium can is 100% recyclable. Every time a can is recycled, enough energy is saved to power a television for about three hours (compared to mining and producing a new can).

Glass

Glass bottles and jars are gathered via curbside collection schemes and bottle banks, where the glass may be sorted into color categories. The collected glass *cullet* is taken to a glass recycling plant where it is monitored for purity and contaminants are removed. The cullet is crushed and added to a raw material mix in a melting furnace. It is then mechanically blown or molded into new jars or bottles. Glass cullet is also used in the construction industry for aggregate and glassphalt. Glassphalt is a road-laying material which comprises around 30% recycled glass. Glass can be recycled indefinitely as its structure does not deteriorate when reprocessed.

Paper

Recycled paper is made from waste paper, usually mixed with fresh wood pulp. If the paper contains ink, it must be deinked. This also removes fillers, clays, and fiber fragments.

Almost all paper can be recycled today, but some types are harder to recycle than others. Papers coated with plastic or aluminum foil, and papers that are waxed, pasted, or gummed are usually not recycled because the process is too expensive. Gift wrap paper also cannot be recycled. Different types of paper are usually sorted before recycling, such as newspapers and cardboard boxes.

Different grades of paper are recycled into different types of new products. Old newspapers are usually made into new newsprint, egg cartons, or paperboard. Old corrugated boxes are made into new corrugated boxes or paperboard. High-grade white office paper can be made into almost any new paper product: stationery, newsprint, magazines, or books.

Sometimes recyclers ask for the removal of the glossy inserts from newspapers because they are a different type of paper. Glossy inserts have a heavy clay coating that some paper mills cannot accept. Most of the clay is removed from the recycled pulp as sludge which must be disposed. If the coated paper is 20% by weight clay, then each ton of glossy paper produces more than 200 kg of sludge and less than 800 kg of fiber. Uncoated (no clay), recycled newsprint gives less sludge and more useable fiber.

Paper can only be recycled a finite number of times due to the shortening of paper fibers making the material less versatile. Often it will be mixed with a quantity of virgin material, referred to as *downcycling*. This does not however exclude the material from being used in other processes such as composting or anaerobic digestion, where further value can be extracted from the material in the form of compost or biogas.

Plastic

Plastic recycling is the process of recovering scrap or waste plastics and reprocessing the material into useful products. Compared to glass or metallic materials, plastic poses unique challenges - because of the massive number of types of plastic, they each carry a resin identification code, and must be sorted before they can be recycled. This can be costly - while metals can be sorted using electromagnets, no such 'easy sorting' capability exists for plastics. In addition to this, while labels do not need to be removed from bottles for recycling, lids are often made from a different kind of non-recyclable plastic.

Plastics recycling rates lag far behind those of other items, such as newspaper and aluminium; consumers are typically unsure of how to recycle plastics, and compared to paper and metals fewer recycling facilities exist.

Textiles

When considering textile recycling one must understand what the material consists of Most textiles are composites of cotton (biodegradable material) and synthetic plastics. The textile's composition will affect its durability and method of recycling.

Workers sort and separate collected textiles into good quality clothing and shoes which can be reused or worn. There is a trend of moving these facilities from developed countries to developing countries.

Damaged textiles are further sorted into grades to make industrial wiping cloths and for use in paper manufacture or material suitable for fibre reclamation and filling products. If textile reprocessors receive wet or soiled clothes however, these may still be disposed of in a landfill, as the washing and drying facilities are not present at sorting units.

Fibre reclamation mills sort textiles according to fibre type and colour. Colour sorting eliminates the need to re-dye the recycled textiles. The textiles are shredded into "shoddy" fibres and blended with other selected fibres, depending on the intended end use of the recycled yarn. The blended mixture is carded to clean and mix the fibres and spun ready for weaving or knitting. The fibres can also be compressed for mattress production. Textiles sent to the flocking industry are shredded to make filling material for car insulation, roofing felts, loudspeaker cones, panel linings and furniture padding.

Timber

Recycling timber has become popular due to its image as an environmentally friendly product, with consumers commonly believing that by purchasing recycled wood the demand for *green timber* will fall and ultimately benefit the environment. Greenpeace also view recycled timber as an environmentally friendly product, citing it as the most preferable timber source on their website. The arrival of recycled timber as a construction product has been important in both raising industry and consumer awareness towards deforestation and promoting timber mills to adopt more environmentally friendly practices.

Other Techniques

Several other materials are also commonly recycled, frequently at an industrial level.

Ship breaking is one example that has associated environmental, health, and safety risks for the area where the operation takes place; balancing all these considerations is an environmental justice problem.

Tires are also commonly recycled. Used tires can be added to asphalt, producing road surfaces that are more durable, create less traffic noise, and absorb precipitation better than traditional asphalt. Tires can also be used to make rubber mulch, which is used on playgrounds for safety.

Metal scavenged from automobiles can also be recycled at an industrial scale.

International Universal Recycling Codes

The communication and identification are laid out in International Universal Recycling Codes. These codes outline what material an item is made from, to facilitate easier reprocessing.

WASTE MANAGEMENT

Waste management is the collection, transport, processing, recycling or disposal of waste materials. The term usually relates to materials produced by human activity, and is generally undertaken to reduce their effect on health, aesthetics or amenity. Waste management is also carried out to reduce the materials' effect on the environment and to recover resources from them. Waste management can involve solid, liquid or gaseous substances, with different methods and fields of expertise for each.

Waste management practices differ for developed and developing nations, for urban and rural areas, and for residential and industrial, producers. Management for non-hazardous residential and institutional waste in metropolitan areas is usually the responsibility of local government authorities, while management for non-hazardous commercial and industrial waste is usually the responsibility of the generator.

Waste Management Methods

Waste management methods vary widely between areas for many reasons, including type of waste material, nearby land uses, and the area available.

The analysis of existing methods of processing waste

Disposal

Landfill

Disposing of waste in a landfill involves burying waste to dispose of it, and this remains a common practice in most countries. Historically, landfills were often established in disused quarries, mining voids or borrow pits. A properly-designed and well-managed landfill can be a hygienic and relatively inexpensive method of disposing of waste materials. Older, poorly-designed or poorly-managed landfills can create a number of adverse environmental impacts such as wind-blown litter, attraction of vermin, and generation of liquid leachate. Another common byproduct of landfills is gas (mostly composed of methane and carbon dioxide), which is produced as organic waste breaks down anaerobically. This gas can create odor problems, kill surface vegetation, and is a greenhouse gas.

Design characteristics of a modern landfill include methods to contain leachate such as clay or plastic lining material. Deposited waste is normally compacted to increase its density and stability, and covered to prevent attracting vermin (such as mice or rats). Many landfills also have landfill gas extraction systems installed to extract the landfill gas. Gas is pumped out of the landfill using perforated pipes and flared off or burnt in a gas engine to generate electricity.

Many local authorities, especially in rural areas, have found it difficult to establish new landfills due to opposition from owners of adjacent land. As a result, solid waste disposal in these areas must be transported further for disposal or managed by other methods. This fact, as well as growing concern about the environmental impacts of excessive materials consumption, has given rise to efforts to minimize the amount of waste sent to landfill in many areas. These efforts include taxing or levying waste sent to landfill, recycling waste products, converting waste to energy, and designing products that use less material.

Incineration

Incineration is a disposal method that involves combustion of waste material. Incineration and other high temperature waste treatment systems are sometimes described as "thermal treatment". Incinerators convert waste materials into heat, gas, steam, and ash.

Incineration is carried out both on a small scale by individuals, and on a large scale by industry. It is used to dispose of solid, liquid and gaseous waste. It is recognised as a practical method of disposing of certain hazardous waste materials (such as biological medical waste). Although it remains a controversial method of waste disposal in many places due to issues such as emission of gaseous pollutants, modern combustion technologies such as the RCBC (rotary cascading bed combustor) allow efficient energy production with emissions well within regulatory standards.

Incineration is common in countries such as Japan where land is more scarce, as these facilities generally do not require as much area as landfills. Waste-to-energy (WtE) or energy-from-waste (EfW) are broad terms for incinerator facilities that burn waste in a furnace or boiler to generate heat, steam and/or electricity.

Recycling

The process of extracting resources or value from waste is generally referred to as recycling, meaning to recover or reuse the material. There are a number of different methods by which waste material is recycled: the raw materials may be extracted and reprocessed, or the calorific content of the waste may be converted to electricity. New methods of recycling are being developed continuously, and are described briefly below.

Physical Reprocessing

The popular meaning of 'recycling' in most developed countries refers to the widespread collection and reuse of everyday waste materials such as empty beverage containers. These are collected and sorted into common types so that the raw materials from which the items are made can be reprocessed into new products. Material for recycling may be collected separately from general waste using dedicated bins and collection vehicles, or sorted directly from mixed waste streams.

The most common consumer products recycled include aluminium beverage cans, steel food and aerosol cans, HDPE and PET bottles, glass bottles and jars, paperboard cartons, newspapers, magazines, and cardboard. Other types of plastic (PVC, LDPE, PP, and PS: see resin identification code) are also recyclable, although these are not as commonly collected. These items are usually composed of a single type of material, making them relatively easy to recycle into new products. The recycling of complex products (such as computers and electronic equipment) is more difficult, due to the additional dismantling and separation required.

Biological Processing

Waste materials that are organic in nature, such as plant material, food scraps, and paper products, can be recycled using biological composting and digestion processes to decompose the organic matter. The resulting organic material is then recycled as mulch or compost for agricultural or landscaping purposes. In addition, waste gas from the process (such as methane) can be captured and used for generating electricity. The intention of biological processing in waste management is to control and accelerate the natural process of decomposition of organic matter.

There are a large variety of composting and digestion methods and technologies varying in complexity from simple home compost heaps, to industrial-scale enclosed-vessel digestion of mixed domestic waste (see Mechanical biological treatment). Methods of biological decomposition are differentiated as being aerobic or anaerobic methods, though hybrids of the two methods also exist.

An example of waste management through composting is the Green Bin Program in Toronto, Canada, where household organic waste (such as kitchen scraps and plant cuttings) are collected in a dedicated container and then composted.

Energy Recovery

The energy content of waste products can be harnessed directly by using them as a direct combustion fuel, or indirectly by processing them into another type of fuel. Recycling through thermal treatment ranges from using waste as a fuel source for cooking or heating, to fuel for boilers to generate steam and electricity in a turbine. Pyrolysis and gasification are two related forms of thermal treatment where waste materials are heated to high temperatures with limited oxygen availability. The process typically occurs in a sealed vessel under high pressure. Pyrolysis of solid waste converts the material into solid, liquid and gas products. The liquid and gas can be burnt to produce energy or refined into other products. The solid residue (char) can be further refined into products such as activated carbon. Gasification is used to convert organic materials directly into a synthetic gas (syngas) composed of carbon monoxide and hydrogen. The gas is then burnt to produce electricity and steam.

Avoidance and Reduction

Another important method of waste management is the prevention of waste material being created. Methods of avoidance include reuse of second-hand products, repairing broken items instead of buying new, designing products to be refillable or reusable (such as cotton instead of plastic shopping bags), encouraging consumers to avoid using disposable products (such as disposable cutlery), and designing products that use less material to achieve the same purpose (for example, lightweighting of beverage cans).

Waste Handling and Transport

Waste collection methods vary widely between different countries and regions. Domestic waste collection services are often provided by local government authorities, or by private industry. Some areas, especially those in less developed countries, do not have a formal waste-collection system.

For example, in Australia most urban domestic households have a 240-litre (63.4 U.S. gallon) bin that is emptied weekly from the curb using side- or rear-loading compactor trucks. In Europe and a few other places around the world, a few communities use a proprietary collection system known as Envac, which conveys refuse via underground conduits using a vacuum system. In Canadian urban centres curbside collection is the most common method of disposal, whereby the city collects waste and/or recyclables and/or organics on a scheduled basis. In rural areas people often dispose of their waste by hauling it to a transfer station. Waste collected is then transported to a regional landfill.

Waste Management Concepts

There are a number of concepts about waste management which vary in their usage between countries or regions. This section presents some of the most general, widely-used concepts.

Waste Hierarchy

The waste hierarchy refers to the "3 Rs" reduce, reuse and recycle, which classify waste management strategies according to their desirability in terms of waste minimization. The waste hierarchy remains the cornerstone of most waste minimization strategies. The aim of the waste hierarchy is to extract the maximum practical benefits from products and to generate the minimum amount of waste.

Extended Producer Responsibility

Extended Producer Responsibility (EPR) is a strategy designed to promote the integration of all costs associated with products throughout their life cycle (including end-of-life disposal costs) into the market price of the product. Extended producer responsibility is meant to impose accountability over the entire lifecycle of products and packaging introduced to the market. This means that firms which manufacture, import and/or sell products are required to be responsible for the products after their useful life as well as during manufacture.

Polluter Pays Principle

The Polluter Pays Principle is a principle where the polluting party pays for the impact caused to the natural environment. With respect to waste management, this generally refers to the requirement for a waste generator to pay for appropriate disposal of the waste.

Waste Management Trade Associations

- Chartered Institute of Wastes Management (UK)
- International Solid Waste Association (International association)
- Solid Waste Association of North America (North American peak industry body)
- Waste Management Association of Australia (Australian peak industry body)
- Saint Lucia Solid Waste Management Authority

WATER PURIFICATION

Water purification is the process of removing contaminants from a raw water source. The goal is to produce water for a specific purpose with a treatment profile designed to limit the inclusion of specific materials; most water is purified for human consumption (drinking water). Water purification may also be designed for a variety of other purposes, including to meet the requirements of medical, pharmacology, chemical and industrial applications. Methods include, but are not limited to: ultra violet light, filtration, water softening, reverse osmosis, ultrafiltration, molecular stripping, deionization, and carbon treatment.

Water purification may remove: particulate sand; suspended particles of organic materal; parasites, Giardia; Cryptosporidium; bacteria; algae; virus; fungi; etc. Minerals calcium, silica, magnesium, etc. and toxic metals (lead, copper, chromium etc). Some purification may be elective in the purification process, including smell (hydrogen sulfide remediation), taste (mineral extraction), and appearance (iron incapsulation).

Governments usually dictate the standards for drinking water quality. These standards will require minimum / maximum set points of contaminants and the inclusion of control elements that produce drinking water. Quality standards in many countries require specific amounts of disinfectant (such as chlorine or ozone) in the water after it leaves the water treatment plant (WTP), to reduce the risk of re-contamination while the water is in the distribution system.

Ground water (usually supplied as well water) is typically a more economical choice than surface water (from rivers, lakes and streams) as a source for drinking, as it is inherently pre-filtered by the aquifer from which it is extracted. Over large areas of the world, aquifers are recharged as part of the hydrologic cycle. In more arid regions, water from an aquifer will have a limited output and can take thousands of years to recharge. Surface water is locally more abundant where subsurface formations do not function as aquifers; however, ground water is far more abundant than the more-visible surface water. Surface water is a typical raw water source used to make drinking water where it is abundant and where ground water is unavailable or of poor quality. However, it is much more exposed to human activity and its byproducts. As a water source it is carefully monitored for the presence of a variety of contaminants by the WTP operators.

It is not possible to tell whether water is safe to drink just by looking at it. Simple procedures such as boiling or the use of a household activated carbon filter are not sufficient for treating all the possible contaminants that may be present in water from an unknown source. Even natural spring water - considered safe for all practical purposes in the 1800s - must now be tested before determining what kind of treatment, if any, is needed. Chemical analysis, while expensive, is the only way to obtain the information necessary for deciding on method of purification.

Sources of Drinking Water

The water emerging from some deep ground water may have fallen as rain many decades, hundreds, thousands or in some cases millions of years ago. Soil and rock layers naturally filter the ground water to a high degree of clarity before it is pumped to the treatment plant. Such water may emerge as springs, artesian springs, or may be extracted from boreholes or wells. Deep ground water is generally of very high bacteriological quality (i.e., pathogenic bacteria such as *Campylobacter* or the pathogenic protozoa *Cryptosporidium* and *Giardia* are typically absent), but the water typically is rich in dissolved solids, especially carbonates and sulfates of calcium and magnesium. Depending on the strata through which the water has flowed, other ions may also be present including chloride, and bicarbonate. There may be a requirement to reduce the iron or manganese content of this water to make it pleasant for drinking, cooking, and laundry use. Disinfection may also be required. Where groundwater recharge is practised, it is equivalent to lowland surface waters for treatment purposes.

1. Shallow groundwaters: Water emerging from shallow groundwaters is usually abstracted from wells or boreholes. The bacteriological quality can be variable depending on the nature of the catchment. A variety of soluble materials may be present including (rarely) potentially toxic metals such as zinc, copper or arsenic. Arsenic contamination of groundwater is a serious problem in some areas, notably from shallow wells in Bangladesh and West Bengal in the Ganges Delta.
2. Upland lakes and reservoirs: Typically located in the headwaters of river

systems, upland reservoirs are usually sited above any human habitation and may be surrounded by a protective zone to restrict the opportunities for contamination. Bacteria and pathogen levels are usually low, but some bacteria, protozoa or algae will be present. Where uplands are forested or peaty, humic acids can color the water. Many upland sources have low pH which require adjustment.

3. Rivers, canals and low land reservoirs: Low land surface waters will have a significant bacterial load and may also contain algae, suspended solids and a variety of dissolved constituents.
4. Atmospheric water generation is a new technology that can provide high quality drinking water by extracting water from the air by cooling the air and thus condensing water vapour.
5. Rainwater harvesting or fog collection which collect water from the atmosphere can be used especially in areas with significant dry seasons and in areas which experience fog even when there is little rain.

Pre-treatment

1. Pumping and containment - The majority of water must be pumped from its source or directed into pipes or holding tanks. To avoid adding contaminants to the water, this physical infrastructure must be made from appropriate materials and constructed so that accidental contamination does not occur.
2. Screening - The first step in purifying surface water is to remove large debris such as sticks, leaves, trash and other large particles which may interfere with subsequent purification steps. Most deep groundwater does not need screening before other purification steps.
3. Storage - Water from rivers may also be stored in bankside reservoirs for periods between a few days and many months to allow natural biological purification to take place. This is especially important if treatment is by slow sand filters. Storage reservoirs also provide a buffer against short periods of drought or to allow water supply to be maintained during transitory pollution incidents in the source river.
4. Pre-conditioning - Many waters rich in hardness salts are treated with soda-ash (Sodium carbonate) to precipitate calcium carbonate out utilising the common ion effect.
5. Pre-chlorination - In many plants the incoming water was chlorinated to minimise the growth of fouling organisms on the pipe-work and tanks. Because of the potential adverse quality effects (see chlorine below), this has largely been discontinued.

Widely varied techniques are available to remove the fine solids, micro-organisms and some dissolved inorganic and organic materials. The choice of method will depend on the quality of the water being treated, the cost of the treatment process and the quality standards expected of the processed water.

pH Adjustment

Distilled water has an average pH of 7 (neither alkaline nor acidic) and sea water has an average pH of 8.3 (slightly alkaline). If the water is acidic (lower than 7), lime or soda ash is added to raise the pH. Lime is the more common of the two additives because it is cheap, but it also adds to the resulting water hardness. Making the water slightly alkaline ensures that coagulation and flocculation processes work effectively and also helps to minimize the risk of lead being dissolved from lead pipes and lead solder in pipe fittings.

Flocculation is a process which starts with clarifying the water. Clarifying means removing any turbidity or color so that the water is clear and colorless. Clarification is done by causing a precipitate to form in the water. Initially the precipitate forms as very small particles but as the water is gently stirred, these particles stick together to form bigger particles - this process is sometimes called flocculation. Many of the small particles that were originally present in the raw water absorb onto the surface of these small precipitate particles and so get incorporated into the larger particles that coagulation produces. In this way the coagulated precipitate takes most of the suspended matter out of the water and is then filtered off, generally by passing the mixture through a coarse sand filter or sometimes through a mixture of sand and granulated anthracite (high quality coal). Anthracite, with its high carbon content, is able to absorb much of the organic matter present in solution and this can remove odour and taste from the water. A precipitate that is widely used to clarify water is iron (III) hydroxide. This is formed first by adjusting (if necessary) the pH of the incoming water to above 7 (by adding lime or sodium hydroxide), then by adding a solution of an iron (III) compound such as iron (III) chloride. Iron (III) hydroxide is extremely insoluble and forms even at a pH as low as 7. Aluminium hydroxide is also widely used as the flocculating precipitate.

Sedimentation

Water exiting the flocculation basin may enter the sedimentation basin, also called a clarifier or settling basin. It is a large tank with slow flow, allowing floc to settle to the bottom. The sedimentation basin is best located close to the flocculation basin so the transit between does not permit settlement or floc break up. Sedimentation basins can be in the shape of a rectangle, where water flows from end to end, or circular where flow is from the center outward. Sedimentation basin outflow is typically over a weir so only a thin top layer-furthest from the sediment-exits.The amount of floc that settles out of the water is dependent on the time the water spends in the basin and the depth of the basin. The retention time of the water must therefore be balanced against the cost of a larger basin. The minimum clarifier retention time is normally 4 hours. A deep basin will allow more floc to settle out than a shallow basin. This is because large particles settle faster than smaller ones, so large particles bump into and integrate smaller particles as they settle. In effect, large particles sweep vertically though the basin and clean out smaller particles on their way to the bottom.

As particles settle to the bottom of the basin a layer of sludge is formed on the floor of the tank. This layer of sludge must be removed and treated. The amount of sludge that is generated is significant, often 3%-5% of the total volume of water that is treated. The cost of treating and disposing of the sludge can be a significant part of the operating cost of a water treatment plant. The tank may be equipped with mechanical cleaning devices that continually clean the bottom of the tank or the tank can be taken out of service when the bottom needs to be cleaned.

Filtration

After separating most floc, the water is filtered as the final step to remove remaining suspended particles and unsettled floc. The most common type of filter is a rapid sand filter. Water moves vertically through sand which often has a layer of activated carbon or anthracite coal above the sand. The top layer removes organic compounds, which contribute to taste and odour. The space between sand particles is larger than the smallest suspended particles, so simple filtration is not enough. Most particles pass through surface layers but are trapped in pore spaces or adhere to sand particles. Effective filtration extends into the depth of the filter. This property of the filter is key to its operation: if the top layer of sand were to block all the particles, the filter would quickly clog.

To clean the filter, water is passed quickly upward through the filter, opposite the normal direction (called *backflushing* or *backwashing*) to remove embedded particles. Prior to this, compressed air may be blown up through the bottom of the filter to break up the compacted filter media to aid the backwashing process; this is known as *air scouring*. This contaminated water can be disposed of, along with the sludge from the sedimentation basin, or it can be recycled by mixing with the raw water entering the plant.

Some water treatment plants employ pressure filters. These work on the same principle as rapid gravity filters, differing in that the filter medium is enclosed in a steel vessel and the water is forced through it under pressure.

Advantages

Filters out much smaller particles than paper and sand filters can.

Filters out virtually all particles larger than their specified pore sizes.

They are quite thin and so liquids flow through them fairly rapidly.

They are reasonably strong and so can withstand pressure differences across them of typically 2-5 atmospheres.

They can be cleaned (back flushed) and reused.

Membrane filters are widely used for filtering both drinking water and sewage (for reuse). For drinking water, membrane filters can remove virtually all particles larger than 0.2 um—including Giardia and cryptosporidium. Membrane filters are an effective form of tertiary treatment when it is desired to reuse the water for industry, for limited domestic purposes, or before discharging the water into a river that is used by towns further downstream. They are widely used in industry, particularly for beverage preparation (including bottled water). However no filtration can remove substances that are actually dissolved in the water such as phosphorus, nitrates and heavy metal ions.

Slow Sand Filters

Slow sand filters may be used where there is sufficient land and space as the water must be passed very slowly through the filters. These filters rely on biological treatment processes for their action rather than physical filtration. The filters are carefully constructed using graded layers of sand with the coarsest at the top and finest at the base. Drains at the base convey treated water away for disinfection. Filtration depends on the development of a thin biological layer on the surface of the filter. An effective slow sand filter may remain in service for many weeks or even months if the pre-treatment is well designed and produces water with a very low available nutrient level which physical methods of treatment rarely achieve. Very low nutrient levels allow water to be safely sent through distribution system with very low disinfectant levels thereby reducing consumer irritation over offensive levels of chlorine and chlorine by-products. Slow sand filters are not backwashed; they are maintained by having the top layer of sand scraped off when flow is eventually obstructed by biological growth. [citation needed]

Ultrafiltration

Ultrafiltration membranes are a relatively new development; they use polymer film with chemically formed microscopic pores that can be used in place of granular media to filter water effectively without coagulants. The type of membrane media determines how much pressure is needed to drive the water through and what sizes of micro-organisms can be filtered out.

Disinfection

Disinfection is accomplished both by filtering out harmful microbes and also by adding disinfectant chemicals in the last step in purifying drinking water. Water is disinfected to kill any pathogens which pass through the filters. Possible pathogens include viruses, bacteria, including *Escherichia coli*, *Campylobacter* and *Shigella*, and protozoans, including *G. lamblia* and other Cryptosporidia. In most developed countries, public water supplies are required to maintain a residual disinfecting agent throughout the distribution system, in which water may remain for days before reaching the consumer. Following the introduction of any chemical disinfecting agent, the water is usually held in temporary storage - often called a contact tank or clear well to allow the disinfecting action to complete.

1. ***Chlorination-*** The most common disinfection method is some form of chlorine or its compounds such as chloramine or chlorine dioxide. Chlorine is a strong oxidant that rapidly kills many harmful micro-organisms.

   Because chlorine is a toxic gas, there is a danger of a release associated with its use. This problem is avoided by the use of sodium hypochlorite, which is a relatively inexpensive solution that releases free chlorine when dissolved in water. Chlorine solutions can be generated on site by electrolyzing common salt solutions. A solid form , calcium hypochlorite exists that releases chlorine on contact with water. Handling the solid, however, requires greater routine human contact through opening bags and pouring than the use of gas cylinders or bleach which are more easily automated. The generation of liquid sodium hypochlorite is both inexpensive and safer than the use of gas or solid chlorine. All forms of chlorine are widely used despite their respective drawbacks. One drawback is that chlorine from any source reacts with natural organic compounds in the water to form potentially harmful chemical by-products trihalomethanes (THMs) and haloacetic acids (HAAs), both of which are carcinogenic in large quantities and regulated by the United States Environmental Protection Agency (EPA). The formation of THMs and haloacetic acids may be minimized by effective removal of as many organics from the water as possible prior to chlorine addition. Although chlorine is effective in killing bacteria, it has limited effectiveness against protozoans that form cysts in water. (*Giardia lamblia* and *Cryptosporidium*, both of which are pathogenic).

2. Chlorine dioxide is another faster-acting disinfectant. It is, however, relatively rarely used, because in some circumstances it may create excessive amounts of chlorite, which is a by-product regulated to low allowable levels in the USA. Chlorine dioxide is made in water and added/used in water to avoid

gas handling problems; chlorine dioxide gas accumulations may spontaneously detonate.

3. *Chloramines* are another chlorine-based disinfectant. Although chloramine is not as strong of an oxidant it does provide a longer lastinge residual than free chlorine, and it not form THMs or haloacetic acids. It is possible to convert chlorine to chloramine by adding ammonia to the water after addition of chlorine: The chlorine and ammonia react to form chloramine. Water distribution systems disinfected with chloramines may experience nitrification, wherein ammonia is used a nutrioent for bacterial growth, with nitrates being generated as a byproduct.
4. *Ozone* (O_3) is a relatively unstable molecule "free radical" of oxygen which readily gives up one atom of oxygen providing a powerful oxidising agent which is toxic to most water borne organisms. It is a very strong, broad spectrum disinfectant that is widely used in Europe. It is an effective method to inactivate harmful protozoans that form cysts. It also works well against almost all other pathogens. Ozone is made by passing oxygen through ultraviolet light or a "cold" electrical discharge. To use ozone as a disinfectant, it must be created on site and added to the water by bubble contact. Some of the advantages of ozone include the production of relatively fewer dangerous by-products (in comparison to chlorination) and the lack of taste and odor produced by ozonation. Although fewer by-products are formed by ozonation, it has been discovered that the use of ozone produces a small amount of the suspected carcinogen Bromate, although little Bromine should be present in treated water. Another one of the main disadvantages of ozone is that it leaves no disinfectant residual in the water. Ozone has been used in drinking water plants since 1906 where the first industrial ozonation plant was built in Nice, France. The U.S. Food and Drug Administration has accepted ozone as being safe; and it is applied as an anti-microbiological agent for the treatment, storage, and processing of foods.
5. *UV radiation (light)* is very effective at inactivating cysts, as long as the water has a low level of colour so the UV can pass through without being absorbed. The main disadvantage to the use of UV radiation is that, like ozone treatment, it leaves no residual disinfectant in the water.

Because neither ozone nor UV radiation leaves a residual disinfectant in the water, it is sometimes necessary to add a residual disinfectant after they are used. This is often done through the addition of chloramines, discussed above as a primary disinfectant. When used in this manner, chloramines provide an effective residual disinfectant with very little of the negative aspects of chlorination.

Additional Treatment Options

1. *Fluoridation-* in many areas fluoride is added to water for the purpose of preventing tooth decay. This process is referred to as water fluoridation. Fluoride is usually added after the disinfection process. In the United States, fluoridation is usually accomplished by the addition of hexafluorosilicic acid, which decomposes in water, yielding fluoride ions.

2. *Water conditioning:* This is a method of reducing the effects of hard water. Hardness salts are deposited in water systems subject to heating because the decomposition of bicarbonate ions creates carbonate ions which crystalise out of the saturated solution of calcium or magnesium carbonate. Water with high concentrations of hardness salts can be treated with soda ash (sodium carbonate) which precipitates out the excess salts, through the Common-ion effect, producing calcium carbonate of very high purity. The precipitated calcium carbonate is traditionally sold to the manufacturers of toothpaste. Several other methods of industrial and residential water treatment are claimed (without general scientific acceptance) to include the use of magnetic or/and electrical fields reducing the effects of hard water. [*citation needed*]

3. *Plumbosolvency reduction:* In areas with naturally acidic waters of low conductivity (i.e surface rainfall in upland mountains of igneous rocks), the water may be capable of dissolving lead from any lead pipes that it is carried in. The addition of small quantities of phosphate ion and increasing the pH slightly both assist in greatly reducing plumbo-solvency by creating insoluble lead salts on the inner surfaces of the pipes.

4. *Radium Removal:* Some groundwater sources contain radium, a radioactive chemical element. Typical sources include many groundwater sources north of the Illinois River in Illinois. Radium can be removed by ion exchange, or by water conditioning. The back flush or sludge that is produced is, however, a low-level radioactive waste.

5. *Fluoride Removal:* Although fluoride is added to water in many areas, some areas of the world have excessive levels of natural fluoride in the source water. Excessive levels can be toxic or cause undesirable cosmetic effects such as staining of teeth. One method of reducing fluoride levels is through treatment with activated alumina.

Other Water Purification Techniques

Other popular methods for purifying water, especially for local private supplies are listed below. In some countries some of these methods are also used for large scale municipal supplies. Particularly important are distillation (de-salination of seawater) and reverse osmosis.

1. Boiling: Water is heated hot enough and long enough to inactivate or kill micro-organisms that normally live in water at room temperature. Near sea level, a vigorous rolling boil for at least one minute is sufficient. At high altitudes (greater than two kilometers or 5000 feet) three minutes is recommended.[1] In areas where the water is "hard" (that is, containing significant dissolved calcium salts), boiling decomposes the bicarbonate ions, resulting in partial precipitation as calcium carbonate. This is the "fur" that builds up on kettle elements, etc., in hard water areas. With the exception of calcium, boiling does not remove solutes of higher boiling point than water and in fact increases their concentration (due to some water being lost as vapour). Boiling does not leave a residual disinfectant in the water. Therefore, water that has been boiled and then stored for any length of time may have acquired new pathogens.
2. Carbon filtering: Charcoal, a form of carbon with a high surface area, absorbs many compounds including some toxic compounds. Water passing through activated charcoal is common in household water filters and fish tanks. Household filters for drinking water sometimes contain silver to release silver ions which have an anti-bacterial effect.
3. Distillation involves boiling the water to produce water vapour. The vapour contacts a cool surface where it condenses as a liquid. Because the solutes are not normally vaporised, they remain in the boiling solution. Even distillation does not completely purify water, because of contaminants with similar boiling points and droplets of unvaporised liquid carried with the steam. However, 99.9% pure water can be obtained by distillation. Distillation does not confer any residual disinfectant and the distillation apparatus may be the ideal place to harbour Legionnaires' disease.
4. Reverse osmosis: Mechanical pressure is applied to an impure solution to force pure water through a semi-permeable membrane. Reverse osmosis is theoretically the most thorough method of large scale water purification available, although perfect semi-permeable membranes are difficult to create. Unless membranes are well-maintained, algae and other life forms can colonise the membranes.
5. Ion exchange: Most common ion exchange systems use a zeolite resin bed to replace unwanted Ca^{2+} and Mg^{2+} ions with benign (soap friendly) Na^{+} or K^{+} ions. This is the common water softener.
6. Electrodeionization: Water is passed between a positive electrode and a

negative electrode. Ion selective membranes allow the positive ions to separate from the water toward the negative electrode and the negative ions toward the positive electrode. High purity deionized water results. The water is usually passed through a reverse osmosis unit first to remove non-ionic organic contaminants.

7. The use of iron in removing arsenic from water. See Arsenic contamination of groundwater#Water purification solutions.
8. Direct contact membrane distillation (DCMD). Applicable to desalination. Heated seawater is passed along the surface of a hydrophobic polymer membrane. Evaporated water passes from the hot side through pores in the membrane into a stream of cold pure water on the other side. The difference in vapour pressure between the hot and cold side helps to push water molecules through.

Portable Water Purification

Portable techniques for purifying water are used for hiking, camping etc. or for use in rural areas or emergency situations. Common techniques include boiling, disinfection with tablets or ultra-filtration using a small hand pump.

Water Purification for Hydrogen Production

For the small scale production of hydrogen, water purifiers are installed to prevent formation of minerals on the surface of the electrodes and to remove organics and chlorine from utility water. First, the water passes through a 20 micrometre interference (mesh or screen filter) filter to remove sand and dust particles, then a charcoal filter using activated carbon to remove organics and chlorine and, finally, a de-ionizing filter to remove metallic ions. Testing can be done before and after the filter to verify the proper removal of barium, calcium, potassium, magnesium, sodium and silicon.

Another method that is used is reverse osmosis.

Safety and Controversies

There has been controversy over the fluoridation of water since the 1930s, when the benefits to dental health were first identified.

Water supplies have sometimes been the subject of concerns over terrorism or actual terrorist threats.

Accidents have also been known to happen. In April, 2007, the water supply of Spencer, Massachusetts became contaminated with excess sodium hydroxide (lye) when its treatment equipment malfunctioned.

INDUSTRIAL WASTEWATER TREATMENT

Industrial wastewater treatment covers the mechanisms and processes used to treat waters that have been contaminated in some way by man's industrial or commercial activities prior to its release into the environment or its re-use.

Most industries produce some wet waste although recent trends in the developed world have been to minimise such production or recycle such waste within the production process. However, many industries remain dependent on processes that produce wastewaters.

Sources of Industrial Wastewater

Agricultural Waste

Iron and Steel Industry

The production of iron from its ores involves powerful reduction reactions in blast furnaces. Cooling waters are inevitably contaminated with products especially ammonia and cyanide. Production of coke from coal in coking plants also requires water cooling and the use of water in by-products separation. Contamination of waste streams includes gasification products such as benzene, naphthalene, anthracene, cyanide, ammonia, phenols , cresols together with a range of more complex organic compounds known collectively as polycyclic aromatic hydrocarbons (PAH).

The conversion of iron or steel into sheet, wire or rods requires hot and cold mechanical transformation stages frequently employing water as a lubricant and coolant. Contaminants include hydraulic oils, tallow and particulate solids. Final treatment of iron and steel products before onward sale into manufacturing includes *pickling* in strong mineral acid to remove rust and prepare the surface for tin or chromium plating or for other surface treatments such as galvanisation or painting. The two acids commonly used are hydrochloric acid and sulphuric acid. Wastewaters include acidic rinse waters together with waste acid. Although many plants operate acid recovery plants, (particularly those using Hydrochloric acid), where the mineral acid is boiled away from the iron salts, there remains a large volume of highly acid ferrous sulphate or ferrous chloride to be disposed of. Many steel industry wastewaters are contaminated by hydraulic oil also known as *soluble oil*

Mines and Quarries

The principal waste-waters associated with mines and quarries are slurries of rock particles in water. These arise from rainfall washing exposed surfaces and haul roads and also from rock washing and grading processes. Volumes of water can be very high, especially rainfall related arisings on large sites. Some specialist separation operations such as coal washing to separate coal from native rock using density gradients can produce wastewater contaminated by fine particulate haematite and surfactants. Oils and hydraulic oils are also common contaminants. Wastewater from metal mines and ore recovery plants are inevitably contaminated by the minerals present in the native rock formations. Following crushing and extraction of the desirable materials, undesirable materials may become contaminated in the wastewater. For metal mines, this can include unwanted metals such as zinc and other materials such as arsenic. Extraction of high value metals such as gold and silver may generate slimes containing very fine particles in where physical removal of contaminants becomes particularly difficult.

Food Industry

Wastewater generated from agricultural and food operations has distinctive characteristics that set it apart from common municipal wastewater managed by public or private wastewater treatment plants throughout the world: it is biodegradable and nontoxic, but that has high concentrations of biochemical oxygen demand (BOD) and suspended solids (SS). The constituents of food and agriculture wastewater are often complex to predict due to the differences in BOD and pH in effluents from vegetable, fruit, and meat products and due to the seasonal nature of food processing and postharvesting.

Processing of food from raw materials requires large volumes of high grade water. Vegetable washing generates waters with high loads of particulate matter and some dissolved organics. It may also contain surfactants.

Animal slaughter and processing produces very strong organic waste from body fluids, such as blood , and gut contents. This wastewater is frequently contaminated by significant levels of antibiotics and growth hormones from the animals and by a variety of pesticides used to control external parasites. Insecticide residues in fleeces is a particular problem in treating waters generated in wool processing.

Processing food for sale produces wastes generated from cooking which are often rich in plant organic material and may also contain salt, flavourings, colouring material and acids or alkali. Very significant quantities of oil or fats may also be present.

Complex Organic Chemicals Industry

A range of industries manufacture or use complex organic chemicals. These include pesticides, Pharmaceuticals, paints and dyes, petro-chemicals, detergents, plastics etc. Waste waters can be contaminated by feed-stock materials, by-products, product material in soluble or particulate form , washing and cleaning agents, solvents and added value products such as plasticisers.

Nuclear Industry

The waste production from the nuclear and radio-chemicals industry is dealt with at Radioactive waste treatment

Water Treatment

Water treatment for the production of drinking water is dealt with elsewhere. Many industries have a need to treat water to obtain very high quality water for demanding purposes. Water treatment produces organic and mineral sludges from filtration and sedimentation. Ion exchange using natural or synthetic resins removes calcium, magnesium and carbonate ions from water, replacing them with hydrogen and hydroxyl ions. Regeneration of ion exchange columns with strong acids and alkalis produces a wastewater rich in hardness ions which are readily precipitated out, especially when in admixture with other wastewaters.

Treatment of Industrial Wastewater

The different types of contamination of wastewater require a variety of strategies to remove the contamination.

Solids Removal

Most solids can be removed using simple sedimentation techniques with the solids recovered as slurry or sludge. Very fine solids and solids with densities close to the density of water pose special problems. In such case filtration or ultra-filtration may be required. Alternatively, flocculation may be used using alum salts or the addition of poly-electrolytes

Oils and Grease Removal

Many oils can be recovered from open water surfaces by skimming devices. However, hydraulic oils and the majority of oils that have degraded to any extent will also have a soluble or emulsified component that will require further treatment to eliminate. Dissolving or emulsifying oil using surfactants or solvents usually exacerbates the problem rather than solving it, producing a very difficult to treat wastewater.

The wastewaters from large-scale industries such as oil refineries, petrochemical plants, chemical plants, and natural gas processing plants commonly contain gross amounts of oil and suspended solids. Those industries use a device known as an API oil-water separator which is designed to separate the oil and suspended solids from their wastewater effluents. The name is derived from the fact that such separators are designed according to standards published by the American Petroleum Institute (API).

The API separator is a gravity separation device designed by using Stokes Law to define the rise velocity of oil droplets based on their density and size. The design is based on the specific gravity difference between the oil and the wastewater because that difference is much smaller than the specific gravity difference between the suspended solids and water. The suspended solids settles to the bottom of the separator as a sediment layer, the oil rises to top of the separator and the cleansed wastewater is the middle layer between the oil layer and the solids.

Typically, the oil layer is skimmed off and subsequently re-processed or disposed of, and the bottom sediment layer is removed by a chain and flight scraper (or similar device) and a sludge pump. The water layer is sent to further treatment consisting usually of a dissolved air flotation (DAF) unit for additional removal of any residual oil and then to some type of biological treatment unit for removal of undesirable dissolved chemical compounds.

Parallel plate separators are similar to API separators but they include tilted parallel plate assemblies (also known as parallel packs). The parallel plates provide more surface for suspended oil droplets to coalesce into larger globules. Such separators still depend upon the specific gravity between the suspended oil and the water. However, the parallel plates enhance the degree of oil-water separation. The result is that a parallel plate separator requires significantly less space than a conventional API separator to achieve the same degree of separation.

Removal of Biodegradable Organics

Biodegradable organic material of plant or animal origin is usually possible to treat using extended conventional wastewater treatment processes such as activated sludge or trickling filter. Problems can arise if the wastewater is excessively diluted with washing water or is highly concentrated such as neat blood or milk. The presence of cleaning agents, disinfectants, pesticides, or antibiotics can have detrimental impacts on treatment processes.

Activated Sludge Process

Activated sludge is a biochemical process for treating sewage and industrial wastewater that uses air (or oxygen) and microorganisms to biologically oxidize organic pollutants, producing a waste sludge (or floc) containing the oxidized material. In general, an activated sludge process includes:

- An aeration tank where air (or oxygen) is injected and thoroughly mixed into the wastewater.
- A settling tank (usually referred to as a "clarifier" or "settler") to allow the waste sludge to settle. Part of the waste sludge is recycled to the aeration

tank and the remaining waste sludge is removed for further treatment and ultimate disposal.

Trickling Filter Process

A trickling filter consists of a bed of rocks, gravel, slag, peat moss, or plastic media over which wastewater flows downward and contacts a layer (or film) of microbial slime covering the bed media. Aerobic conditions are maintained by forced air flowing through the bed or by natural convection of air. The process involves adsorption of organic compounds in the wastewater by the microbial slime layer, diffusion of air into the slime layer to provide the oxygen required for the biochemical oxidation of the organic compounds. The end products include carbon dioxide gas, water and other products of the oxidation. As the slime layer thickens, it becomes difficult for the air to penetrate the layer and an inner anaerobic layer is formed.

The components of a complete trickling filter system are: fundamental components:

- A bed of filter medium upon which a layer of microbial slime is promoted and developed.
- An enclosure or a container which houses the bed of filter medium.
- A system for distributing the flow of wastewater over the filter medium.
- A system for removing and disposing of any sludge from the treated effluent.

The treatment of sewage or other wastewater with trickling filters is among the oldest and most well characterized treatment technologies.

A trickling filter is also often called a *trickle filter*, *trickling biofilter*, *biofilter*, *biological filter* or *biological trickling filter*.

Treatment of Other Organics

Synthetic organic materials including solvents, paints, pharmaceuticals, pesticides, coking products and so forth can be very difficult to treat. Treatment methods are often specific to the material being treated. Methods include distillation, adsorption, vitrification, incineration, chemical immobilisation or landfill disposal. Some materials such as some detergents may be capable of biological degradation and in such cases, a modified form of wastewater treatment can be used.

Treatment of Acids and Alkalis

Acids and alkalis can usually be neutralised under controlled conditions. Neutralisation frequently produces a precipitate that will require treatment as a solid residue that may also be toxic. In some cases, gasses may be evolved requiring treatment for the gas stream. Some other forms of treatment are usually required following neutralisation.

Waste streams rich in hardness ions as from de-ionisation processes can readily loose the hardness ions in a buildup of precipitated calcium and magnesium salts. This precipitation process can cause severe *furring* of pipes and can, in extreme cases, cause the blockage of disposal pipes. A 1 metre diameter industrial marine discharge pipe serving a major chemicals complex was blocked by such salts in the 1970s. Treatment is by concentration of de-ionisation waste waters and disposal to landfill or by careful pH management of the released wastewater.

Treatment of Toxic Materials

Toxic materials including many organic materials, metals (such as zinc, silver, cadmium, thallium etc.) acids, alkalis, non-metallic elements (such as arsenic or selenium) are generally resistant to biological processes unless very dilute. Metals can often be precipitated out by changing the pH or by treatment with other chemicals. Many, however, are resistant to treatment or mitigation and may require concentration followed by landfilling or recycling.

LIST OF WASTE WATER TREATMENT TECHNOLOGIES

The following page consist of a list of waste water treatment technologies:

Environmental Science

- Activated sludge
- Aerated lagoon
- Aerobic granular reactor
- Anaerobic clarigester
- Anaerobic digestion
- API oil-water separator
- Anaerobic lagoon
- Belt press
- Bioconversion of biomass to mixed alcohol fuels
- Biolytix
- Carbon filtering
- Cesspit
- Chlorine disinfection

- Combined sewer
- Composting toilet
- Constructed wetland
- Dissolved air flotation
- Distillation
- Electrocoagulation
- Electrodeionization
- Electrolysis
- Expanded granular sludge bed digestion
- Facilitative lagoon
- Flocculation & sedimentation
- Flotation process
- Froth flotation
- Humanure (composting)
- Imhoff tank
- Iodine
- Ion exchange
- Living machines
- Membrane Bioreactor
- Nanotechnology
- N-Viro
- Parallel plate oil-water separator
- Reed bed
- Reverse osmosis
- Rotating biological contactor
- Septic tank
- Sequencing batch reactor
- Sewage treatment
- Trickling filter
- Ultraviolet disinfection

- Upflow anaerobic sludge blanket digestion
- Wet oxidation

Environmental Technology

- Air pollution control
- Air pollution dispersion modeling
- Alternative energy
- Biofuel
- Composting
- Conservation biology
- Conservation ethic
- Ecoforestry
- Energy conservation
- Energy development
- Environmental design
- Environmental impact assessment
- Environmental preservation
- Green building
- Green computing
- Hydrogen technologies
- Industrial wastewater treatment
- Natural building
- Recycling
- Renewable energy
- Renewable energy development
- Remediation
- Solid waste treatment
- Sustainable architecture
- Sustainable energy
- Sustainable development
- Waste water treatment

- Water purification
- Waste management

LIST OF SOLID WASTE TREATMENT TECHNOLOGIES

The following page contains a list of different forms of solid waste treatment technologies and facilities employed in waste management infrastructure.

Waste Handling Facilities

- Civic amenity site (CA Site)
- Transfer station

Established Waste Treatment Technologies

- Composting
- Incineration
- Landfill
- Windrow composting

Advanced Waste Treatment Technologies

Also called alternative waste treatment technologies

- Anaerobic digestion
- Alcohol/ethanol production
  - Bioconversion of biomass to mixed alcohol fuels (pilot scale)
- Biodrying
- Gasification
- In-vessel composting
- Mechanical biological treatment
- Mechanical heat treatment
- Plasma arc waste disposal (pilot scale)
- Pyrolysis
- Recycling
- Sewage treatment
- Tunnel composting
- UASB (applied to solid wastes)

- Waste autoclave
- Waste-to-energy

REFERENCES

Masters, Gilbert M. Introduction to Environmental Engineering. 2nd ed. Upper Saddle River, NJ: Prentice Hall, 1998.

Viessman, Warren, and Mark J. Hammer. Water Supply and Pollution Control. 7th ed. Upper Saddle River, NJ: Prentice Hall, 2005.

Tchobanoglous, G., Burton, F.L., and Stensel, H.D. (2003). *Wastewater Engineering (Treatment Disposal Reuse) / Metcalf & Eddy, Inc.*, 4th Edition, McGraw-Hill Book Company.

Beychok, Milton R. (1967). *Aqueous Wastes from Petroleum and Petrochemical Plants*, 1st Edition, John Wiley & Sons. LCCN 67019834.

American Petroleum Institute (API) (February 1990). *Management of Water Discharges: Design and Operations of Oil-Water Separators*, 1st Edition, American Petroleum Institute.

Beychok, Milton R. (December 1971). "Wastewater treatment". *Hydrocarbon Processing*: 109-112.

Glossary

Abiotic factors, n: Environmental influences produced other than by living organisms; for example, temperature, wind patterns, humidity, pH, substrate rock type, and other physical and chemical influences.

Abiotic, adj: Nonliving or not containing any living organisms.

Absolute poverty, n: The lack of sufficient income in cash or exchange items for meeting the most basic needs of food, clothing, and shelter.

Acid fallout, n: Molecules of acid formed from reactions high in the atmosphere involving nitrogen, sulfur oxides, and water vapor that settle out of the atmosphere without any additional water.

Acid precipitation, n: Includes acid rain, acid fog, acid snow, and any other form of precipitation that is more acidic that normal (i.e., less that pH 5.6). Excess acidity is derived from certain air pollutants, namely sulfur dioxide and oxides of nitrogen. The effects can include: fish kills and *eutrophication* of lakes; tree kills, leading to soil erosion; and physical corrosive damage to vehicles and buildings. Many historic buildings in Europe and the NE United States are suffering damage from severe corrosion due to acid precipitation.

Aerobe, n: An organism that utilizes atmospheric oxygen (0 *2*) in its metabolic pathways. An organism that must have oxygen in order to survive is an *obligate aerobe*.

Aerobic, adj: Living or occurring only in the presence of oxygen: aerobic bacteria. 2. Of or relating to aerobes, organisms that require and utilize oxygen. 3. Involving or improving oxygen consumption by the body: aerobic exercise.

Agroforestry, n: Production of tree crops in a manner similar to agriculture. Also, production of trees along with regular crops.

Anaerobe, n.: 1: An organism capable of living in the absence of free oxygen (O2). 2: *Obligate anaerobe*: An organism that **must** live without oxygen, for whom oxygen (O2) is toxic.

Anaerobic, adj.: 1: Lacking or seriously depleted of oxygen. Opposite of aerobic. 2: Of or relating to organisms, such as certain bacteria, that can live in the absence of atmospheric oxygen (indeed, for most anaerobic bacteria, oxygen is toxic).

Antibody, A protein produced by certain white blood cells in response to a foreign substance (antigen). Each antibody can bind only to a specific antigen.

Antigen, Any foreign or "non-self" substance that, when introduced into the body, causes the immune system to create an antibody.

Autotroph, n: Literally, "self eater." Organisms capable of producing their own food. See *primary producers* . Contrast with *heterotroph* .

Background extinction rate, n: Normal rate of extinction — as a natural part of the evolutionary process — of various species as a result of changes in local environmental conditions and the actions of natural evolutionary forces. Extinctions not caused or contributed to by the actions of humans.

Bacteriophage lambda, A virus which infects *E. coli* , and which is often used in molecular genetics experiments as a vector, or cloning vehicle. Recombinant phages can be made in which certain non-essential l DNA is removed and replaced with the DNA of interest. The phage can accommodate a DNA "insert" of about 15-20 kb. Replication of that virus will thus replicate the investigator's DNA. One would use phage l rather than a plasmid if the desired piece of DNA is rather large.

Base, Alkaline chemical substance, in particular the cyclic nitrogen compounds found in DNA and RNA.

Base-pairing, When two complementary bases (A with T or G with C) recognize each other and are held together by hydrogen bonds.

Binding site, A place on cellular DNA to which a protein (such as a transcription factor) can bind. Typically, binding sites might be found in the vicinity of genes, and would be involved in activating transcription of that gene (promoter elements), in enhancing the transcription of that gene (enhancer elements), or in reducing the transcription of that gene (silencers). NOTE that whether the protein in fact performs these functions may depend on some condition, such as the presence of a hormone, or the tissue in which the gene is being examined. Binding sites could also be involved in the regulation of chromosome structure or of DNA replication.

Binomial nomenclature, n: The two-name system, developed by Carolus Linnaeus (the founder of modern taxonomy), used to assign scientific names to all living things. Homo sapiens, for example, is the scientific name for humans. The first name is the genus name and is always capitalized. This is sort of like your last name... it belongs to several of your close relatives, too, and it shows that you are all closely related. The second name is the species name is always lower case. This is like your first name, which no one else in your circle of relatives posseses and so it uniquely identifies you. Memory tool: you probably know the meanings of the terms generic (i.e. general, broad) and specific (i.e. precise, exact). These terms come from the same origins as genus and species, so recalling their meaning will help you recall the relationship between the two portions of a scientific name.

Bioaccumulation, n: An increase in the concentration of a chemical in specific organs or tissues at a level higher than would normally be expected.

Biodegradable, adj: Able to be broken down into simpler substances (elements and compounds) by naturally occuring decomposers. Essentially, anything that can be ingested by an organism without causing that organism harm. 2. Nontoxic and able to be decomposed in relatively short period even on a *human time scale* .

Biodiversity, n: The variety of biotic factors found within a specified geographic region. 2. The combined differences of living things, generally classified in four broad categories: (i) **Genetic Diversity**: Variety among individuals within a species — or, more specifically, the variety in the DNA of a species. See also "alleles." (ii) **Species Diversity**: Variety of different organisms at the species taxonomic level. See also *species* and *taxonomy* . (iii) **Cultural Diversity**: Variety of learned behaviors among individuals of a species. (iv) **Ecosystem Diversity**: Variety of biomes and habitats occuring in the biosphere.

Biome, n: A specific type of terrestrial region inhabited by well-defined types of life, especially zones of vegetation, that generally cannot live outside that specific region. Examples include types of deserts ("high desert" like the Mojave or "low desert" like the Chihuahua), grasslands (prairies, coastal dunes), and forests (lodgepole pine vs. taiga; temperate rain forest; bamboo forest, tropical rain forest, cloud forest, etc.).

Biosphere, n: The portion of the earth and its atmosphere in which living organisms exist or that is capable of supporting life. 2. All of earth's ecosystems combined into one inclusive unit. Also called the "ecosphere." 3. The living organisms and their environment composing the biosphere. "...all life on earth and the realms that support it, from the outermost reaches of the atmosphere to the deepest trenches of the seas." National Geographic Atlas of the World, 6th Edition.

Biotechnology, A set of biological techniques developed through basic research and now applied to research and product development.

Biotin, A colorless crystalline vitamin, $C_{10}H_{16}N_2O_3S$, of the vitamin B complex, essential for the activity of many enzyme systems and found in large quantities in liver, egg yolk, milk, and yeast. Also known as vitamin H.

Blotting, A technique for detecting one RNA within a mixture of RNAs (a Northern blot) or one type of DNA within a mixture of DNAs (a Southern blot). A blot can prove whether that one species of RNA or DNA is present, how much is there, and its approximate size. Basically, blotting involves gel electrophoresis, transfer to a blotting membrane (typically nitrocellulose or activated nylon), and incubating with a radioactive probe. Exposing the membrane to X-ray film produces darkening at a spot correlating with the position of the DNA or RNA of interest. The darker the spot, the more nucleic acid was present there.

Carrying capacity, n: The amount of animal or plant life (or industry) that can be supported indefinitely on available resources; the number of individuals that the resources of a habitat can support. Also called biological carrying capacity.

Cell, Small, watery, membrane-bound compartment filled with chemicals; the basic unit of any living thing.

Chromosome walking, A technique for cloning everything in the genome around a known piece of DNA (the starting probe). You screen a genomic library for all clones hybridizing with the probe, and then figure out which one extends furthest into the surrounding DNA. The most distal piece of this most distal clone is then used as a probe, so that ever more distal regions can be cloned. This has been used to move as much as 200 kb away from a given starting point (an immense undertaking). Typically used to "walk" from a starting point towards some nearby gene in order to clone that gene. Also used to obtain the remainder of a gene when you have isolated a part of it.

Clone (verb), To "clone" something is to produce copies of it. To clone a piece of DNA, one would insert it into some type of vector (say, a plasmid) and put the resultant construct into a host (usually a bacterium) so that the plasmid and insert replicate with the host. An individual bacterium is isolated and grown and the plasmid containing the "cloned" DNA is re-isolated from the bacteria, at which point there will be many millions of copies of the DNA - essentially an unlimited supply. Actually, an investigator wishing to clone some gene or cDNA rarely has that DNA in a purified form, so practically speaking, to "clone" something involves screening a cDNA or genomic library for the desired clone. See also "Probe" for a description of how one might start a cloning project, and "Screening" for how the probe in used.

Consensus sequence, A 'nominal' sequence inferred from multiple, imperfect examples. Multiple lanes of shotgun sequence can be merged to show a consensus sequence. The optimal sequence of nucleotides recognized by some factor. A DNA binding site for a protein may vary substantially, but one can infer the consensus sequence for the binding site by comparing numerous examples. For example, the (fictitious) transcription factor ZQ1 usually binds to the sequences AAAGTT, AAGGTT or AAGATT. The consensus sequence for that factor is said to be AARRTT (where R is any purine, i.e. A or G). ZQ1 may also be able to weakly bind to ACAGTT (which differs by one base from the consensus).

Conservation biology, n: Multidisciplinary science created to deal with the crisis of maintaining the genes, species, communities, and ecosystems that make up earth's biological diversity. Its goals are to investigate human impacts on biodiversity and to develop practical approaches to preserving biodiversity and ecological integrity.

Conservation-tillage farming, n: Crop cultivation in which the soil is disturbed little (minimum-tillage farming) or not at all (no-till farming) to reduce soil erosion, lower labor costs, and save energy.

Contig, Several uses, all nouns. The term comes from a shortening of the word 'contiguous'. A 'contig' may refer to a map showing placement of a set of clones that completely, contiguously cover some segment of DNA in which you are interested. Also called the 'minimal tiling path'. More often, the term 'contig' is used to refer to the final product of a shotgun sequencing project. When individual lanes of sequence information are merged to infer the sequence of the larger DNA piece, the product consensus sequence is called a 'contig'.

Coral bleaching, n: The loss of color from a coral as it expels its zooxanthellae-usually a stress response.

Cosmid, A type of vector used for cloning 35-45 kb of DNA. These are plasmids carrying a phage l cos site (which allows packaging into l capsids), an origin of replication and an antibiotic resistance gene. A plasmid of 40 kb is very difficult to put into bacteria, but can replicate once there. Cosmids, however, have a cos site, and thus can be packaged into l phage heads (a reaction which can be performed *in vitro*) to allow efficient introduction into bacteria (you'll have to look up the cos site elsewhere).

Cost-benefit analysis, n: Estimates and comparison of short-term and long-term costs (losses) and benefits (gains) from an economic decision. If the estimated benefits exceed the estimated costs, the decision to buy an economic good or provide a public good is considered worthwhile.

Debt-for-nature swap, n: Agreement in which a certain amount of foreign debt is canceled in exchange for local currency investments that will improve natural resource management or protect certain areas in the debtor country from harmful development.

Deforestation, n: Removal of trees from a forested area without adequate replanting.

Demographic transition, n: Hypothesis that countries, as they become industrialized, have declines in death rates followed by declines in birth rates.

Denature, To destroy the 3-D folding of a protein or other polymeric molecule.

Deoxyribonuclease, An enzyme that cuts DNA into shorter pieces.

Deoxyribose, The sugar with five carbon atoms that is found in DNA.

Desertification, n: Conversion of rangeland, rain-fed cropland to desert-like land, with a drop in agricultural productivity of 10% or more. It is usually caused by a combination of overgrazing, soil erosion, prolonged drought, and climate change.

Dioxin, n: A synthetic, organic chemical of the chlorinated hydrocarbon class. It is one of the most toxic compounds known to humans, having many harmful effects, including induction of cancer and birth defects, even in extremely minute concentrations. It has become a widespread environmental pollutant because of the use of certain herbicides that contain dioxin as a contaminant.

DNA Cloning, Facilitates isolation and manipulation of fragments of an organism's genome by replicating them as part of an independent vector.

DNA Double Helix, Most common form of DNA. Two separate and antiparallel chains of DNA are wound around each other in a right - handed helical path, with sugar - phosphate backbones on the outside and bases connected by hydrogen bonds on the inside. A (Adenine) pairs with T (Thymine); G (Guanine) pairs with C (Cytosine). The two chains are complementary.

DNA Libraries, Consist of sets of random cloned fragments of either genomic or cDNA, each in a separate vector molecule. Are used in isolation of unknown genes.

DNA Sequencing, The widely used Sanger's enzymic method uses dideoxynucleotide's as chain terminators to produce a ladder of molecule's generated by polymerase extension of a primer. Determines the sequences of the bases in DNA.

DNA, The substance of heredity; a linear molecule that carries the genetic information that cells need to replicate and to produce proteins and DNA. DNA stands for deoxyribonucleic acid.

DNase, Deoxyribonuclease, a class of enzymes which digest DNA. The most common is DNase I, an endonuclease which digests both single and double-stranded DNA.

Dot blot, A technique for measuring the amount of one specific DNA or RNA in a complex mixture. The samples are spotted onto a hybridization membrane (such as nitrocellulose or activated nylon, etc.), fixed and hybridized with a radioactive probe. The extent of labeling (as determined by autoradiography and densitometry) is proportional to the concentration of the target molecule in the sample. Standards provide a means of calibrating the results.

Double-helix, Structure in which two strands of DNA are twisted spirally around each other.

Ecological efficiency, n: The percentage of energy in biomass produced by one trophic level that is incorporated into biomass by the next highest trophic level.

Ecological fitness, n: The number of a parent's young that live to reproduce; divided by two if sexual reproduction is involved.

Ecological succession, n: Process in which communities of plant and animal species in a particular area are replaced over time by a series of different and often more complex communities.

Ecologically sustainable development, n: Development in which the total human population size and resource use in the world (or in a region) are limited to a level that does not exceed the carrying capacity or the existing natural capital and is therefore sustainable.

Ecologist, n: A scientist who studies *ecology* .

Ecology, n: The study of the relationships between organisms and their environments, including: the interactions of living organisms with one another and with their non-living surroundings, the flow of matter and energy in an environment, and the structure and functions of nature. Also called *bionomics* . 2. The relationship between organisms and their environment. 3. The branch of sociology that is concerned with studying the relationships between human groups and their physical and social environments. Also called *human ecology* . 4. The study of the detrimental effects of modern civilization on the environment, with a view toward prevention or reversal through conservation. A component of the field of *human ecology* .

Ecosystem services , n: Services, vital to the support of human life, provided by intact natual ecosystems. These include the purification of air and water, detoxification and dècomposition of wastes, regulation of climate, regeneration of soil fertility, and production and maintenance of biodiversity, from which key ingredients of our agricultural, pharmaceutical, and industrial enterprises are derived. **Historically, the nature and value of Earth's life support systems have largely been ignored until their disruption or loss highlighted their importance**. Read more about the value of ecosystem services and how they are measured.

Ecosystem, n: An ecological community of various plants, animals, and other organisms, interacting with each other and with the nonliving resources in their environment, all functioning as a unit.

Ecotourism, n: The enterprises involved in promoting tourism of unusual or interesting ecological sites. Environmentally, culturally, and scientifically responsible tourism that takes great efforts to ensure tourism revenues benefit the local communities where tourism occurs, the local inhabitants benefit the most economically (revenues are not returned to the traveler's country of origin) and native culture is not diluted with imported tourist cultures. Ecotourism safeguards the nature of the attraction that instigated the tourism and serves to strengthen conservation and scientific research efforts in the area. Very few large corporations who claim to engage in ecotourism actually do so. The most notorious and damaging of tourism industries — the cruise line industry — is an excellent example of a branch of travel that claims to be environmentally-friendly but is in fact extremely damaging, both culturally and ecologically.

El Nino Southern Oscillation (ENSO), n: Flip-flopping pressure systems in the South Pacific that trigger short-lived global changes in climate. Warm waters from the western Pacific move across the ocean, just below the equator, and significantly warm the eastern tropical Pacific.

Electrophoresis, Movement of charged molecules towards an electrode of the opposite charge; used to separate nucleic acids and proteins.

Emergent, n: A tree with a canopy that forms about the general upper most continuous canopy.

Emerging disease, n: The Institute for Medicine defines emerging and re-emerging diseases as: "New, re-emerging, or drug-resistant infections whose incidence in humans has increased in the last two decades or whose incidence threatens to increase in the near future."

Endangered Species Act, n: The United States federal legislation that mandates protection of species and their habitats that are determined by scientifica consensus to be in danger of extinction.

Endangered species, n: Wild species with so few individual survivors that the species could soon become extinct in all or most of its natural range.

Endonuclease, An enzyme which digests nucleic acids starting in the middle of the strand (as opposed to an exonuclease, which must start at an end). Examples include the restriction enzymes, DNase I and RNase A.

Enhancer, An enhancer is a nucleotide sequence to which transcription factor(s) bind, and which increases the transcription of a gene. It is NOT part of a promoter; the basic difference being that an enhancer can be moved around anywhere in the general vicinity of the gene (within several thousand nucleotides on either side or even within an intron), and it will still function. It can even be clipped out and spliced back in backwards, and will still operate. A promoter, on the other hand, is position- and orientation-dependent. Some enhancers are "conditional" - in other words, they enhance transcription only under certain conditions, for example in the presence of a hormone.

Environment, n: All external conditions and factors, living and nonliving (chemicals and energy), that affect an organism or other specified system during its lifetime; the earth's life-support systems for us and for all other forms of life - in effect another term for describing solar capital and earth capital.

Environmental degradation, n: A reduction of an ecosystem's or habitat's ability to support its natural biota. 2. Depletion or destruction of a potentially renewable resource such as soil, grassland, forest, or wildlife by using it at a faster rate than it is naturally replenished. If such use is continued, the resource can become nonrenewable (on a human timescale) or nonexistant. 3. Pollution, toxification, or other alteration of an environment that makes it less productive, hospitable, usable, or enjoyable.

Environmental worldview, n: How individuals think the world works, what they think their role in the world should be, and what they believe is right and wrong environmental behavior (i.e. ethics).

Enzyme, A protein that acts as a catalyst, affecting the rate at which chemical reactions occur in cells.

Epidemiology, n: Study of the patterns of disease or other harmful effects from toxic exposure within defined groups of people to find out why some people get sick and some do not.

ERE, Estrogen Response Element. A binding site in a promoter to which the activated estrogen receptor can bind. The estrogen receptor is essentially a transcription factor which is activated only in the presence of estrogens. The activated receptor will bind to an ERE, and transcription of the adjacent gene will be altered. See also "Response element".

ERID, acronym: Emerging and Re-emerging Infectious Diseases. See *emerging disease*

Ethnobotany, n: The study of indigenous knowledge bases regarding plants and their uses.

Ethnopharmacology, n: The study of indigenous knowledge bases regarding medicines and how they are produced, as well as the medical practices, treatment protocols, etc. that utilize these medicines.

Evolutionary Footprinting, One can infer which portions of a gene are important by comparing the sequence of that gene with its cognates from other species. A plot showing the regions of high conservation will presumably reflect the regions that are functional in all the test species. In theory, the more species involved in the comparison, the more stringent the result can be (i.e. the more the conserved regions will reflect truly important sequences). Care must be taken, however, to use species in which the function of the gene has not diverged excessively, or the outcome will be uninformative.

Exon, Those portions of a genomic DNA sequence which WILL be represented in the final, mature mRNA. The term "exon" can also be used for the equivalent segments in the final RNA. Exons may include coding sequences, the 5' untranslated region or the 3' untranslated region.

Exonuclease, An enzyme which digests nucleic acids starting at one end. An example is Exonuclease III, which digests only double-stranded DNA starting from the 3' end.

Exponential growth, n: Growth in which some quantity, such as population size or economic output, increases by a fixed percentage of the whole in a given time; when the increase in quantity over a long enough time is plotted, this type of growth typically yields a curve shaped like the letter J.

Expression clone, This is a clone (plasmid in a bacteria, or maybe a l phage in bacteria) which is designed to produce a protein from the DNA insert. Mammalian genes do not function in bacteria, so to get bacterial expression from your mammalian cDNA, you would place its coding region (i.e. no introns) immediately adjacent to bacterial transcription/translation control sequences. That artificial construct (the "expression clone") will produce a pseudo-mammalian protein if put back into bacteria. Often, that protein can be recognized by antibodies raised against the authentic mammalian protein, and vice versa.

Expression of a gene, Synthesis of the protein, or the gene product - result of transcription plus processing plus translation.

Expression, To "express" a gene is to cause it to function. A gene which encodes a protein will, when expressed, be transcribed and translated to produce that protein. A gene which encodes an RNA rather than a protein (for example, a rRNA gene) will produce that RNA when expressed.

Extant, adj: A species that is still alive and reproducing. All species that currently live on earth are extant.

Extinct, adj: A species that is no longer living on earth. All representatives of the species are dead. All the species that once occupied the earth but are no longer living are extinct. We know of their existence through studying the *fossil record* . Compare to *extant* .

Extinction, n: Complete disappearance of a species from the earth. This happens when a species cannot adapt and successfully reproduce under new environmental conditions, when it evolves (through a process called *radiation*) into one or more new species, or when every member of the species is killed by overpredation, pollution, or other man-made causes.

First law of human ecology, n: We can never do merely one thing. Any intrusion into nature has numerous effects, many of which are unpredictable. For example, one classic dilemma is the case of behavioral biologists who observe their study subjects at close range: Are the observed behaviors truly natural or are they influenced by the researcher's presence?

Food chain, n: Figure of speech describing the dependence of *heterotrophs* on other organisms for food, progressing in a series beginning with *primary producers* (plants) and ending with the largest carnivores. The food chain is used as a figurative image for educational purposes only... real trophic systems resemble webs rather than chains. See *food web* .

Food web, n: The combination of all the feeding relationships that exist in an ecosystem. Most prey species are eaten by many differet predators, and most predators eat more than one prey item. As a result, a picture of a trophic system with lines (representing ecological relationships) drawn between predators and prey soon resembles an intricate web.

Footprinting, A technique by which one identifies a protein binding site on cellular DNA. The presence of a bound protein prevents DNase from "nicking" that region, which can be detected by an appropriately designed gel.

Fossil fuel, n: Products of partial of complete decomposition of plants and animals that occur as crude oil, coal, natural gas, or heavy oils as a result of exposure to heat and pressure in earth's crust over millions of years.

Fossil record, n: The cumulative taxonomic information and historical perspective provided by the wealth and diversity of fossils and related geologic data stored in the earth's crust.

Fossil, n: A remnant, impression, mineralized mold, amber encasement, or other trace of a once-living organism. Technically, anything that once lived and has been permanently preserved is a fossil, but the most common usage implies great age. This common usage of fossil generally refers to the mineralized remains or impressions, preserved in stone (almost always sedimentary rock), of extinct organisms from past geologic ages.

Gel electrophoresis, Electrophoresis of charged moleculaes thorugh a gel meshwork in order to sort them by size.

Gel shift assay, (aka gel mobility shift assay (GMSA), band shift assay (BSA), electrophoretic mobility shift assay (EMSA)) A method by which one can determine whether a particular protein preparation contains factors which bind to a particular DNA fragment. When a radiolabeled DNA fragment is run on a gel, it shows a characteristic mobility. If it is first incubated with a cellular extract of proteins (or with purified protein), any protein-DNA complexes will migrate slower than the naked DNA - a shifted band.

Gene pool, n: The sum total of all the genes that exist among all the individuals of a species.

Gene therapy, Treatment that alters genes - the basic units of heredity found in all cells in the body.

Gene, A unit of DNA which performs one function. Usually, this is equated with the production of one RNA or one protein. A gene contains coding regions, introns, untranslated regions and control regions.

Gene, A unit of inheritance; a working subunit of DNA. Each of the body's 50,000 to 100,000 genes contains the code for a specific product, typically, a protein such as an gene's coded information is translated into the structures present and operating in the cell (either proteins or RNAs).

Genetic code, The sequence of nucleotides, coded in triplets (codons) along the mRNA, that determines the sequence of amino acids in protein synthesis.

Genetic engineering, n: The artificial transfer of specific genes from one organism to another.

Genome, All the genetic material in the chromosomes of a particular organism; its size is generally given as its total number of base pairs.

Genome, The total DNA contained in each cell of an organism. Mammalian genomic DNA (including that of humans) contains 6x109 base pairs of DNA per diploid cell. There are somewhere in the order of a hundred thousand genes, including coding regions, 5' and 3' untranslated regions, introns, 5' and 3' flanking DNA. Also present in the genome are structural segments such as telomeric and centromeric DNAs and replication origins, and intergenic DNA.

Genomic blot, A type of Southern blot specifically used to analyze a mixture of DNA fragments derived from total genomic DNA. Because genomic DNA is very complicated, when it has been digested with restriction enzymes, it produces a complex set of fragments ranging from tens of bp to tens of thousands of bp. However, any specific gene will be reproducibly found on only one or a few specific fragments. A million identical cells will produce a million identical restriction fragments for any given gene, so probing a genomic Southern with a gene-specific probe will produce a pattern of perhaps one or just a few bands.

Genomic clone, A piece of DNA taken from the genome of a cell or animal, and spliced into a bacteriophage or other cloning vector. A genomic clone may contain coding regions, exons, introns, 5' flanking regions, 5' untranslated regions, 3' flanking regions, 3' untranslated regions, or it may contain none of these...it may only contain intergenic DNA (usually not a desired outcome of a cloning experiment!).

Genomic DNA library, Library made from chromosomal DNa which therefore contains introns.

Genotype, Genetic constitution of an organism; describes an organism at the genetic leve.

Genotype, Two uses: one is a verb, the other a noun. To 'genotype' (verb) is to example polymorphisms (e.g. RFLPs, microsatellites, SNPs) present in a sample of DNA. You might be looking for linkage between a microsatellite marker and an unknown disease gene. With such information, you can infer the chromosomal location of the unknown gene, and can sometimes identify the gene. As a noun, a 'genotype' is the result of a genotyping experiment, be it a SNP or microsat or whatever.

Geologic time scale, n: Occurring at such a slow pace, or at such infrequent intervals, as to be imperceptible to humans. 2. Occurring in a pre-human era. 3. The whole of earth's history, as opposed to the very recent period when humans have walked the earth. One common and effective means of conceptualizing the disparity between the geologic time scale and the *human time scale* is the "calendar year history model," wherein the entire history of the planet is condensed into a single calendar year. In this model, human ancestors do not appear until late December and Homo sapiens does not arise until the last second before midnight on December 31st.

Geology, n: The branch of science that deals with the earth's history, particularly its physical history, as recorded in the substrate and the *fossil record* .

Geopolitics, n: The study of the influence of such factors as geography, natural resources, economics, and demography on the politics (especially the foreign policy) of nations.

Global warming, n: The term given to the possibility that Earth's atmosphere is gradually warming because of the greenhouse effect of carbon dioxide and other gases. Global warming is thought by many to be the most serious global environmental issue facing our society.

GRE, Glucocorticoid Response Element, A binding site in a promoter to which the activated glucocorticoid receptor can bind. The glucocorticoid receptor is essentially a transcription factor which is activated only in the presence of glucocorticoids. The activated receptor will bind to a GRE, and transcription of the adjacent gene will be altered. See also "Response element".

Green Revolution, n: Refers to the development and introduction of new varieties of wheat and rice (mainly) that increased yields per acre dramatically in some countries.

Greenhouse effect, n: A natural effect that traps heat in the atmosphere (troposphere) near the earth's surface. Some of the heat flowing back toward space from the earth's surface is absorbed by water vapor, carbon dioxide, ozone, and several other gases in the lower atmosphere (troposphere) and then radiated back toward the earth's surface. If the atmospheric concentrations of these greenhouse gases rise and are not removed by other natural processes, the average temperature of the lower atmosphere will gradually increase.

Greenhouse gases, n: Gases in the earth's lower atmosphere (troposphere) that cause the greenhouse effect. Examples are carbon dioxide, chlorofluorocarbons, ozone, methane, water vapor, and nitrous oxide.

Gross primary productivity, n: The rate at which an ecosystem's producers capture and store a given amount of chemical energy as biomass in a given length of time.

Habitat, n: Place or type of place where an organism, population, or community lives.

Hazardous waste, n: Any solid, liquid, or containerized gas that can catch fire easily, is corrosive to skin tissue or metals, is unstable and can explode or release toxic fumes, or has harmful concentrations of one or more toxic materials that can leach out.

Helix-loop-helix, A protein structural motif characteristic of certain DNA-binding proteins.

Heterotroph, n: Literally, "eats others." An organism that must consume other organisms to fuel its metabolism. Animals, including humans, are heterotrophs. Heterotrophic, adj.

hnRNA, Heterogeneous nuclear RNA; refers collectively to the variety of RNAs found in the nucleus, including primary transcripts, partially processed RNAs and snRNA. The term hnRNA is often used just for the unprocessed primary transcripts, however.

Host strain (bacterial), The bacterium used to harbor a plasmid. Typical host strains include HB101 (general purpose *E. coli* strain), DH5a (ditto), JM101 and JM109 (suitable for growing M13 phages), XL1-Blue (general-purpose, good for blue/white lacZ screening). Note that the host strain is available in a form with no plasmids (hence you can put one of your own into it), or it may have plasmids present (especially if you put them there). Hundreds, perhaps thousands, of host strains are available.

Human capital, n: Physical and mental talents of people used to produce, distribute, and sell an economic good.

Human ecology, n: See *ecology* .

Human time scale, n: Occurring within a short enough time frame that the event can be perceived, remembered, and recounted by humans through oral traditions, written histories, or other mechanisms of human memory. Compare to *geologic time scale* .

Hybrid, n: The offspring of two parents from separate (though closely related) species. Usually sterile, though occassionally able to breed back into one of the parent lines. A hybrid can almost never produce viable offspring when mated with another hybrid. A common example is a mule, which is produced by breeding a horse with a donkey (note that the horse must be the mother, due to the large size of the foal). Hybridization is fairly common among wind-pollinated plants, while hybridization is quite uncommon among higher animals.

Hybridization, The reaction by which the pairing of complementary strands of nucleic acid occurs. DNA is usually double-stranded, and when the strands are separated they will re-hybridize under the appropriate conditions. Hybrids can form between DNA-DNA, DNA-RNA or RNA-RNA. They can form between a short strand and a long strand containing a region complementary to the short one. Imperfect hybrids can also form, but the more imperfect they are, the less stable they will be (and the less likely to form). To "anneal" two strands is the same as to "hybridize" them.

Hydrologic cycle, n: Biogeochemical cycle that collects, purifies, and distributes the earth's fixed supply of water, from the environment to living organisms and then back to the environment.

Hydrosphere, n: The earth's liquid water (oceans, lakes other bodies of surface water, and underground water), the earth's frozen water (polar ice caps, floating ice caps, and ice in soil known as permafrost), and small amounts of water vapor in the atmosphere.

Indicator species, n: Species that serve as early warnings that a community or an ecosystem is being degraded. Fish and amphibians make particularly excellent indicator species. Large predators (those generally at the apex of the food pyramid) are also good indicators in many habitats.

Insert, In a complete plasmid clone, there are two types of DNA - the "vector" sequences and the "insert". The vector sequences are those regions necessary for propagation, antibiotic resistance, and all those mundane functions necessary for useful cloning. In contrast, however, the insert is the piece of DNA in which you are really interested.

Integrated pest management (IPM), n: Combined use of biological, chemical, and cultivation methods in proper sequence and timing to keep the size of a pest population below the size that causes economically unacceptable loss of a crop or livestock animal.

Intergenic, Between two genes; e.g. intergenic DNA is the DNA found between two genes. The term is often used to mean non-functional DNA (or at least DNA with no known importance to the two genes flanking it). Alternatively, one might speak of the "intergenic distance" between two genes as the number of base pairs from the polyA site of the first gene to the cap site of the second. This usage might therefore include the promoter region of the second gene.

Interspecific competition, n: Members of two or more species trying to use the same limited resources in an ecosystem.

Intraspecific competition, n: Two or individual organisms of a single species trying to use the same limited resources in an ecosystem.

Intron, Introns are portions of genomic DNA which ARE transcribed (and thus present in the primary transcript) but which are later spliced out. They thus are not present in the mature mRNA. Note that although the 3' flanking region is often transcribed, it is removed by endonucleolytic cleavage and not by splicing. It is not an intron.

Intron, Segment of a gene that does not code for protein.

Keystone species, n: Species that play roles affecting many other organisms in an ecosystem.

Kinase, A kinase is in general an enzyme that catalyzes the transfer of a phosphate group from ATP to something else. In molecular biology, it has acquired the more specific verbal usage for the transfer onto DNA of a radiolabeled phosphate group. This would be done in order to use the resultant "hot" DNA as a probe.

Knock-out experiment, A technique for deleting, mutating or otherwise inactivating a gene in a mouse. This laborious method involves transfecting a crippled gene into cultured embryonic stem cells, searching through the thousands of resulting clones for one in which the crippled gene exactly replaced the normal one (by homologous recombination), and inserting that cell back into a mouse blastocyst. The resulting mouse will be chimaeric but, if you are lucky (and if you've gotten this far, you obviously are), its germ cells will carry the deleted gene. A few rounds of careful breeding can then produce progeny in which both copies of the gene are inactivated.

Kwashiorkor, n: Type of malnutrition that occurs in infants and very young children when they are weaned from mother's milk to a starchy diet low in protein.

Land-use planning, n: Process for deciding the best present and future use of each parcel of land in an area.

Late successional plant species, n: Mostly trees that can tolerate shade and that form a relatively stable complex forest community.

Laterite: n. Product of rock decomposition with high iron and aluminum hydroxide content. Generally bright red to deep orange in color. 2. Land, usually in the tropics, baked by the sun after deforestation removes the protective and restorative forest layer above the soil. Abiotic hardpack ground, red in color. Normal soil microbiotic community, as well as macrobiotic flora and fauna, are absent. Prone to extensive erosion due to lack of plant cover. Lateritized hillsides have contributed to several devastating and deadly landslides in tropical countries.

Laterization, n: The process of turning formerly healthy soils into laterite. What becomes of tropical forest lands when deforested and left exposed to the elements. The ground becomes extremely hard and cannot be penetrated by germinating forest seeds, so recolonization is slow or absent.

Leucine zipper, A motif found in certain proteins in which Leu residues are evenly spaced through an a-helical region, such that they would end up on the same face of the helix. Dimers can form between two such proteins. The Leu zipper is important in the function of transcription factors such as Fos and Jun and related proteins.

Library, A library might be either a genomic library, or a cDNA library. In either case, the library is just a tube carrying a mixture of thousands of different clones - bacteria or l phages. Each clone carries an "insert" - the cloned DNA. A cDNA library is usually just a mixture of bacteria, where each bacteria carries a different plasmid. Inserted into the plasmids (one per plasmid) are thousands of different pieces of cDNA (each typ. 500-5000 bp) copied from some source of mRNA, for example, total liver mRNA. The basic idea is that if you have a large enough number of different liver-derived cDNAs carried in those bacteria, there is a 99% probability that a cDNA copy of any given liver mRNA exists somewhere in the tube. The real trick is to find the one you want out of that mess - a process called screening (see "Screening"). A genomic library is similar in concept to a cDNA library, but differs in three major ways - 1) the library carries pieces of genomic DNA (and so contains introns and flanking regions, as well as coding and untranslated); 2) you need bacteriophage l or cosmids, rather than plasmids, because... 3) the inserts are usually 5-15 kb long (in a l library) or 20-40 kb (in a cosmid library). Therefore, a genomic library is most commonly a tube containing a mixture of l phages. Enough different phages must be present in the library so that any given piece of DNA from the source genome has a 99% probability of being present.

Lichen, n: A *symbiotic* relationship between a fungus and a moss. The moss does most of the work, producing sugars for the lichen's collective metabolic pathways. Lichen are generally low-growing, vary in color from bright orange or yellow to gray or black, and are often found growing on rocks and tree bark. An easy mnemonic to assist recall of the nature of a lichen's symbiosis is: "A fungus took a *likin'* to a moss, and now they live together."

Ligase, An enzyme, T4 DNA ligase, which can link pieces of DNA together. The pieces must have compatible ends (both of them blunt, or else mutually compatible sticky ends), and the ligation reaction requires ATP.

Ligation, The process of splicing two pieces of DNA together. In practice, a pool of DNA fragments are treated with ligase (see "Ligase") in the presence of ATP, and all possible splicing products are produced, including circularized forms and end-to-end ligation of 2, 3 or more pieces. Usually, only some of these products are useful, and the investigator must have some way of selecting the desirable ones.

Linker, A small piece of synthetic double-stranded DNA which contains something useful, such as a restriction site. A linker might be ligated onto the end of another piece of DNA to provide a desired restriction site.

M13, A bacteriophage which infects certain strains of *E. coli* . The salient feature of this phage is that it packages only a single strand of DNA into its capsid. If the investigator has inserted some heterologous DNA into the M13 genome, copious quantities of single-stranded DNA can subsequently be isolated from the phage capsids. M13 is often used to generate templates for DNA sequencing.

Marasmus, n: Nutritional-deficiency disease caused by a diet that does not have enough calories and protein to maintain good health.

Mass extinction, n: A catastrophic, widespread — often global — event in which major groups of species are wiped out over a relatively short period when compared to normal (*background*) extinction rates. There have been five major mass extinctions, of natural causes (in at least one case due to an asteroid impacting the earth), in the earth's history. We are now entering a sixth great mass extinction, this time of unnatural causes... human activities.

Median lethal dose (LD50), n: Amount of a toxic material per unit body weight of test animals that kills half the test population in a certain time.

Mitosis, The process of nuclear division in cells that produces daughter cells that are genetically identical to each other and to the parent cell.

Monocropping, v: The act of planting and maintaining a *monoculture* .

Monoculture, n: Cultivation of a single crop, usually on a large area of land. This unnatural agricultural system generally requires the use of large quantities of artificial fertilizers, herbicides, pesticides, nematocides, and other pest control efforts. Even with these chemical aids, monocultures are prone to disease outbreaks and pest infestations, largely due to the genetic homogeneity of such systems.

mRNA, "messenger RNA" or sometimes just "message"; an RNA which contains sequences coding for a protein. The term mRNA is used only for a mature transcript with polyA tail and with all introns removed, rather than the primary transcript in the nucleus. As such, an mRNA will have a 5' untranslated region, a coding region, a 3' untranslated region and (almost always) a poly(A) tail. Typically about 2% of the total cellular RNA is mRNA.

Natural resources, n: Nutrients and minerals in the soil and deeper layers of the earth's crust; water; wild and domestic plants and animals; air; and other resources produced by the earth's natural processes.

Natural selection, n: One of several gradual mechanisms through which evolution occurs. Process by which a particular beneficial gene (or set of genes) is reproduced more than other genes in succeeding generations due to selective pressures in the environment that favor the beneficial gene. The result of natural selection is a population that contains a greater proportion of organisms better adapted to certain environmental conditions.

Negative feedback loop, n: Situation in which a change in a certain direction provides information that causes a system to change less in that direction. This is a common regulatory mechanism and is widely used in animals to control hormone levels in the blood. For example, the hormones that control ovulation in humans are on a negative feedback loop.

Nick translation, A method for incorporating radioactive isotopes (typically 32P) into a piece of DNA. The DNA is randomly nicked by DNase I, and then starting from those nicks DNA polymerase I digests and then replaces a stretch of DNA. Radiolabeled precursor nucleotide triphosphates can thus be incorporated.

Nitrogen cycle, n: Cyclic movement of nitrogen in different chemical forms from the environment to organisms and then back to the environment.

Nitrogen fixation, n: The process of chemically converting nitrogen gas (N_2) from the air into compounds, such as nitrates (NO_3), nitrites (NO_2), or ammonia (NH_3), that can be used by plants in building amino acids and other nitrogen-containing organic molecules.

Nonbiodegradable, adj: Not able to be consumed and/or broken down by biological organisms. Nonbiodegradable substances include plastics, aluminum, and many chemicals used in industry and agriculture. Particularly dangerous are nonbiodegradable chemicals that are also toxic and tend to accumulate in organisms.

Non-coding strand, Anti-sense strand. See "Sense strand" for a discussion of sense strand vs. anti-sense strand.

Nonrenewable resource, n: Resource that exists in a fixed amount (stock) in various places in the earth's crust and has the potential for renewal only by geological, physical, and chemical processes taking place over hundreds of millions to billions of years. Examples are copper, aluminum, coal, and oil. We classify these resources as exhaustible because we are extracting and using them at a much faster rate than they were formed.

Northern blot, A technique for analyzing mixtures of RNA, whereby the presence and rough size of one particular type of RNA (usually an mRNA) can be ascertained. See "Blotting" for more information. After Dr. E. M. Southern invented the Southern blot, it was adapted to RNA and named the "Northern" blot.

NT, Abbreviation for nucleotide; i.e. the monomeric unit from which DNA or RNA are built. One can express the size of a nucleic acid strand in terms of the number of nucleotides in its chain; hence 'nt' can be a measure of chain length.

Nuclear run-on, A method used to estimate the relative rate of transcription of a given gene, as opposed to the steady-state level of the mRNA transcript (which is influenced not just by transcription rates, but by the stability of the RNA). This technique is based on the assumption that a highly-transcribed gene should have more molecules of RNA polymerase bound to it than will the same gene in a less-active state. If properly prepared, isolated nuclei will continue to transcribe genes and incorporate 32P into RNA, but only in those transcripts that were in progress at the time the nuclei were isolated. Once the polymerase molecules complete the transcript they have in progress, they should not be able to re-initiate transcription. If that is true, then the amount of radiolabel incorporated into a specific type of mRNA is theoretically proportional to the number of RNA polymerase complexes present on that gene at the time of isolation. A very difficult technique, rarely applied appropriately from what I understand.

Nuclease, An enzyme which degrades nucleic acids. A nuclease can be DNA-specific (a DNase), RNA-specific (RNase) or non-specific. It may act only on single stranded nucleic acids, or only on double-stranded nucleic acids, or it may be non-specific with respect to strandedness. A nuclease may degrade only from an end (an exonuclease), or may be able to start in the middle of a strand (an endonuclease). To further complicate matters, many enzymes have multiple functions; for example, Bal31 has a 3'-exonuclease activity on double-stranded DNA, and an endonuclease activity specific for single-stranded DNA or RNA.

Nucleotide, A molecule consisting of a nucleoside (purine or pyrimidine base plus a pentose sugar)with one or more phosphate groups; a monomer or subunit of nucleic acid, consisting of a base plus sugar plus 3 phosphates.

Nucleus, The nucleus of a cell is an internal compartment surrounded by the nuclear membrane and containing the chromosomes. Only the cells of higher organisms have nuclei.

Nutrient, n: Any food or element an organism must take in to live, grow, or reproduce. Plant: An essential element in a particular ion or molecule that can be absorbed and used by he plant. For example, carbon, hydrogen, nitrogen, and phosphorus are essential elements; carbon dioxide, water, nitrate (NO *3*), and phosphate (PO *4*) are respective nutrients. Animal: Materials such as protein, vitamins, and minerals that are required for growth, maintenance, and repair of the body and also materials such as carbohydrates that are required for energy.

Old-growth forest, n: Virgin and old, second growth forests containing trees that are often hundreds, sometimes thousands, of years old. These are the richest forest biomes with the widest arrays of niche microhabitats and the broadest biodiversity, especially in the tropics.

Oncogene, A gene in a tumor virus or in cancerous cells which, when transferred into other cells, can cause transformation (note that only certain cells are susceptible to transformation by any one oncogene). Functional oncogenes are not present in normal cells. A normal cell has many "proto-oncogenes" which serve normal functions, and which under the right circumstances can be activated to become oncogenes. The prefix "v-" indicates that a gene is derived from a virus, and is generally an oncogene (like *v-src* , *v-ras*, *v-myb* , etc). See also "Transformation (with respect to cultured cells)".

Open reading frame, Any region of DNA or RNA where a protein could be encoded. In other words, there must be a string of nucleotides (possibly starting with a Met codon) in which one of the three reading frames has no stop codons. See "Reading frame" for a simple example.

Open system, n: A system, such as a living organism, in which both matter and energy are exchanged between the system and the environment.

Optimum sustainable population, n: the number of animals which will result in the maximum productivity of the population or the species, keeping in mind the carrying capacity of the habitat and the health of the ecosystem.

Organelle, A discrete subcellular structure that has a specialised function (eg, nucleus, mitochondrion).

Organic, adj: All living things, and products that are uniquely produced by living things, such as wood, leather, and sugar. 2. All chemical compounds or molecules, natural or synthetic, that contain carbon atoms as an integral part of their structure.

Origin of replication, Nucleotide sequences present in a plasmid which are necessary for that plasmid to replicate in the bacterial host. (Abbr. "ori")

Overburden, n: Layer of soil and rock overlying a mineral deposit, removed during surface mining.

Overconsumption, n: Situation in which some people consume much more than they need at the expense of those who can not meet their basic needs- and at the expense of earth's present and future life-support systems for humans and other forms of life.

Overfishing, n: Harvesting so many fish of a species (especially immature fish) that there is not enough breeding stock left to replenish the species, such that it is not profitable to harvest them, leading to *commercial extinction* .

Overgrazing, n: Destruction of vegetation when too many grazing animals feed too long and exceed the carrying capacity of a rangeland area.

Overnutrition, n: Diet so high in calories, saturated (animal) fats, salt, sugar, and processed foods, and so low in vegetables and fruits that the consumer runs high risks of diabetes, hypertension, heart disease, and other health hazards.

Paleoecology, n: The study of ancient ecosystems. Paleoecologists use data from such sources as tree rings, geologic deposits, fossils (pollen is a particularly popular tool), and coral bores to reconstruct the climate and ecology or ancient ecosystems.

Pathway elucidcation, The term biochemical pathways has principally referred to metabolic pathways, which are the pathways by which a cell converts compounds that enter it into cellular components (*e.g.*, small molecules and macromolecules including proteins, nucleic acids, storage carbohydrates, and fatty acids) and by which the cell derives energy. Signaling pathways are biochemical pathways that regulate cellular characteristics and processes such as physiology, proliferation, changes in shape and motility, differentiation, adhesion, and intercellular interactions. Pathway elucidation refers to the understanding of these.

pBR322, A common plasmid. Along with the obligatory origin of replication, this plasmid has genes which make the *E. coli* host resistant to ampicillin and tetracycline. It also has several restriction sites (BamHI, PstI, EcoRI, HindIII etc.) into which DNA fragments could be spliced in order to clone them.

PCR, see Polymerase Chain Reaction.

Peptide, A molecule consisting of 2 to approximately 20 amino acids connected by peptide bonds; a short segment of a larger protein or a completely functional molecule unto itself.

Phagemid, A type of plasmid which carries within its sequence a bacteriophage replication origin. When the host bacterium is infected with "helper" phage, the phagemid is replicated along with the phage DNA and packaged into phage capsids.

Phosphorus cycle, n: Cyclic movement of phosphorus, in varying chemical forms, from the environment to organisms and then back to the environment.

Pioneer species, n: First hardy, often *xerophytic* , species (often microbes, mosses, and lichens) that begin colonizing a site as the first stage of ecological succession.

Plasmid, A circular piece of DNA present in bacteria or isolated from bacteria. *Escherichia coli*, the usual bacteria in molecular genetics experiments, has a large circular genome, but it will also replicate smaller circular DNAs as long as they have an "origin of replication". Plasmids may also have other DNA inserted by the investigator. A bacterium carrying a plasmid and replicating a million-fold will produce a million identical copies of that plasmid. Common plasmids are pBR322, pGEM, pUC18.

Pollutant, n: A particular chemical or form of energy that can adversely affect the health, survival, or activities of humans or other living organisms.

PolyA tail, After an mRNA is transcribed from a gene, the cell adds a stretch of A residues (typically 50-200) to its 3' end. It is thought that the presence of this "polyA tail" increases the stability of the mRNA (possibly by protecting it from nucleases). Note that not all mRNAs have a polyA tail; the histone mRNAs in particular do not.

Polymerase Chain Reaction (PCR) is the basis for a number of extremely important methods in molecular biology. It can be used to detect and measure vanishingly small amounts of DNA and to create customized pieces of DNA. It has been applied to clinical diagnosis and therapy, to forensics and to vast numbers of research applications. It would be difficult to overstate the importance of PCR to science.

Polymerase chain reaction, A technique for replicating a specific piece of DNA *in-vitro* , even in the presence of excess non-specific DNA. Primers are added (which initiate the copying of each strand) along with nucleotides and Taq polymerase. By cycling the temperature, the target DNA is repetitively denatured and copied. A single copy of the target DNA, even if mixed in with other undesirable DNA, can be amplified to obtain billions of replicates. PCR can be used to amplify RNA sequences if they are first converted to DNA via reverse transcriptase. This two-phase procedure is known as 'RT-PCR'.

Polymerase, An enzyme which links individual nucleotides together into a long strand, using another strand as a template. There are two general types of polymerase — DNA polymerases (which synthesize DNA) and RNA polymerase (which makes RNA). Within these two classes, there are numerous sub-types of polymerase, depending on what type of nucleic acid can function as template and what type of nucleic acid is formed. A DNA-dependant DNA polymerase will copy one DNA strand starting from a primer, and the product will be the complementary DNA strand. A DNA-dependant RNA polymerase will use DNA as a template to synthesize an RNA strand.

Population change, n: An increase or decrease in the size of a population. It is equal to (births + immigration) - (deaths + emigration).

Population density, n: Number of organisms in a particular population found in a specified area.

Population dispersion, n: General pattern in which the members of a population are arranged throughout its habitat.

Population distribution, n: Variation of population density over a particular geographical area. For example, a country has a high population density in its urban areas and a much lower population density in rural areas.

Population, n: A group within a single species, the individuals of which can and do freely interbreed. Breeding between populations of the same species is less common because of differences in location, culture, nationality, and so on.

Positive feedback loop, n: Situation in which a change in a certain direction provides information that causes a system to change further in the same direction. This can lead to a runaway or vicious cycle.

Post-transcriptional regulation, Any process occurring after transcription which affects the amount of protein a gene produces. Includes RNA processing efficiency, RNA stability, translation efficiency, protein stability. For example, the rapid degradation of an mRNA will reduce the amount of protein arising from it. Increasing the rate at which an mRNA is translated will increase the amount of protein product.

Post-translational processing, The reactions which alter a protein's covalent structure, such as phosphorylation, glycosylation or proteolytic cleavage.

Post-translational regulation, Any process which affects the amount of protein produced from a gene, and which occurs AFTER translation in the grand scheme of genetic expression. Actually, this is often just a buzz-word for regulation of the stability of the protein. The more stable a protein is, the more it will accumulate.

Potentially renewable resource, n: Resource that theoretically can last indefinitely without reducing the available supply, either because it is replaced more rapidly through natural processes than are nonrenewable resources or because it is potentially inexhaustible (solar energy). Examples are trees in forests, grasses in grasslands, wild animals, fresh surface water in lakes and streams, most groundwater, fresh air, and fertile soil. If such a resource is used faster than it is replenished, it can be depleted and converted into a nonrenewable resource.

Poverty, n: Inability to meet basic needs for food, clothing, and shelter.

PRE, Progesterone Response Element, A binding site in a promoter to which the activated progesterone receptor can bind. The progesterone receptor is essentially a transcription factor which is activated only in the presence of progesterone. The activated receptor will bind to a PRE, and transcription of the adjacent gene will be altered. See also "Response element".

Primary producer, n: An organism, such as a plant or microbe, that makes its own food and forms the bottom-most tier in a trophic system. Primary producers are the basis of the food web in most ecosystems (the exceptions are *open system* communities based entirely on scavenging nutrients flushed into the system from elsewhere, such as some deep sea communities — though even in these cases, the food flushed into the system comes from another system where primary producers are the basis of the *trophic pyramid*). Primary producers are able to convert abiotic raw materials into biotic tissue, either by capturing the sun's energy through *photosynthesis* (plants) or by harnessing the energy in chemical bonds through *chemosynthesis* (some microbes).

Primary transcript, When a gene is transcribed in the nucleus, the initial product is the primary transcript, an RNA containing copies of all exons and introns. This primary transcript is then processed by the cell to remove the introns, to cleave off unwanted 3' sequence, and to polyadenylate the 5' end. The mature message thus formed is then exported to the cytoplasm for translation.

Primer, A small oligonucleotide (anywhere from 6 to 50 nt long) used to prime DNA synthesis. The DNA polymerases are only able to extend a pre-existing strand along a template; they are not able to take a naked single strand and produce a complementary copy of it *de-novo*. A primer which sticks to the template is therefore used to initiate the replication. Primers are necessary for DNA sequencing and PCR.

Probe, A fragment of DNA or RNA which is labeled in some way (often incorporating 32P or 35S), and which is used to hybridize with the nucleic acid in which you are interested. For example, if you want to quantitate the levels of alpha subunit mRNA in a preparation of pituitary RNA, you might make a radiolabeled RNA *in-vitro* which is complementary to the mRNA, and then use it to probe a Northern blot of the pit RNA. A probe can be radiolabeled, or tagged with another functional group such as biotin. A probe can be cloned DNA, or might be a synthetic DNA strand. As an example of the latter, perhaps you have isolated a protein for which you wish to obtain a cDNA or genomic clone. You might (pay to) microsequence a portion of the protein, deduce the nucleic acid sequence, (pay to) synthesize an oligonucleotide carrying that sequence, radiolabel it and use it as a probe to screen a cDNA library or genomic library. A better way is to call up someone who already has the clone.

Processing, The reactions occurring in the nucleus which convert the primary RNA transcript to a mature mRNA. Processing reactions include capping, splicing and polyadenylation. The term can also refer to the processing of the protein product, including proteolytic cleavages, glycosylation, etc.

Promoter, The first few hundred nucleotides of DNA "upstream" (on the 5' side) of a gene, which control the transcription of that gene. The promoter is part of the 5' flanking DNA, i.e. it is not transcribed into RNA, but without the promoter, the gene is not functional. Note that the definition is a bit hazy as far as the size of the region encompassed, but the "promoter" of a gene starts with the nucleotide immediately upstream from the cap site, and includes binding sites for one or more transcription factors which can not work if moved farther away from the gene.

Protein, A large, complex molecule composed of amino acids. Wide variety of functions *e.g.* enzymes catalysing biochemical reactions, membrane receptors in cell signalling, transport and storage, antibody proteins, nutritional proteins, contractile muscle fibers (actin and myosin) etc.

Proteomics, The identification and analysis of the total protein complement expressed by any given cell type under defined conditions.

Proto-oncogene, A gene present in a normal cell which carries out a normal cellular function, but which can become an oncogene under certain circumstances. The prefix "c-" indicates a cellular gene, and is generally used for proto-oncogenes (examples: *c-myb* , *c-myc* , *c-fos* , *c-jun* , etc).

Pulsed field gel electrophoresis, (PFGE) A gel technique which allows size-separation of very large fragments of DNA, in the range of hundreds of kb to thousands of kb. As in other gel electrophoresis techniques, populations of molecules migrate through the gel at a speed related to their size, producing discrete bands. In normal electrophoresis, DNA fragments greater than a certain size limit all migrate at the same rate through the gel. In PFGE, the electrophoretic voltage is applied alternately along two perpendicular axes, which forces even the larger DNA fragments to separate by size.

Pyramid of biomass, n: Diagram representing the biomass (total dry weight of living organisms) that can be supported at each trophic level in a food web. The bottom of the pyramid is comprised of *primary producers* , while the peak of the pyramid is topped by one (or at most a small handful) *apex predator* . Humans are abnormal in that we cross all ecosystems and biomass pyramids, and in almost every one (excepting the polar caps and deepest of oceanic environments) we are the dominant apex predator.

Pyramid of energy flow, n: Also called a *trophic pyramid* . Diagram representing the flow of energy through each trophic level in a food chain or food web. With each energy transfer, only a small part (typically 10%) of the usable energy entering one trophic level is transferred to the organisms at the next trophic level, with the remaining 90% lost as heat or expended in metabolic processes.

Random primed synthesis, If you have a DNA clone and you want to produce radioactive copies of it, one way is to denature it (separate the strands), then hybridize to that template a mixture of all possible 6-mer oligonucleotides. Those oligos will act as primers for the synthesis of labeled strands by DNA polymerase (in the presence of radiolabeled precursors).

Receptor, A protein or group of associated proteins in a cell or on its surface that selectively binds a specific substance (called a ligand).

Recombination, The natural process of breaking and rejoining DNA strands to produce new combinations of genes and, thus, generate genetic variation.

Repetitive DNA, A surprising portion of any genome consists not of genes or structural elements, but of frequently repeated simple sequences. These may be short repeats just a few nt long, like CACACA etc. They can also range up to a few hundred nt long. Examples of the latter include Alu repeats, LINEs, SINEs. The function of these elements is often unknown. In shorter repeats like di- and tri-nucleotide repeats, the number of repeating units can occasionally change during evolution and descent. They are thus useful markers for familial relationships and have been used in paternity testing, forensic science and in the identification of human remains.

Resource economics, n: The study of natural ecosystem services and the economic values, in terms real-world currencies and capital valuations, of those services. One of the goals of resource economics is to assist policy makers in performing the cost-benefit analysis of various plans of action or inaction with regard to the natural world. The value of an ecosystem service is determined by calculating what it would cost to perform the same service artificially if the naturally-occuring service were disrupted or destroyed.

Resource partitioning, n: Process of dividing up resources in a ecosystem so that species with similar requirements (overlapping ecological niches) use the same scarce resources at different times, in different ways, or in different places.

Response element, By definition, a "response element" is a portion of a gene which must be present in order for that gene to respond to some hormone or other stimulus. Response elements are binding sites for transcription factors. Certain transcription factors are activated by stimuli such as hormones or heat shock. A gene may respond to the presence of that hormone because the gene has in its promoter region a binding site for hormone-activated transcription factor. Example: the glucocorticoid response element (GRE).

Restriction fragment, The piece of DNA released after restriction digestion of plasmids or genomic DNA. See "Restriction enzyme". One can digest a plasmid and isolate one particular restriction fragment (actually a set of identical fragments). The term also describes the fragments detected on a genomic blot which carry the gene of interest.

Restriction, To "restrict" DNA means to cut it with a restriction enzyme. See "Restriction Enzyme".

Reverse transcriptase, An enzyme which will make a DNA copy of an RNA template - a DNA-dependant RNA polymerase. RT is used to make cDNA; one begins by isolating polyadenylated mRNA, providing oligo-dT as a primer, and adding nucleotide triphosphates and RT to copy the RNA into cDNA.

RFLP, Restriction fragment length polymorphism; the acronym is pronounced "riflip". Although two individuals of the same species have almost identical genomes, they will always differ at a few nucleotides. Some of these differences will produce new restriction sites (or remove them), and thus the banding pattern seen on a genomic Southern will thus be affected. For any given probe (or gene), it is often possible to test different restriction enzymes until you find one which gives a pattern difference between two individuals - a RFLP. The less related the individuals, the more divergent their DNA sequences are and the more likely you are to find a RFLP.

Riboprobe, A strand of RNA synthesized *in-vitro* (usually radiolabeled) and used as a probe for hybridization reactions. An RNA probe can be synthesized at very high specific activity, is single stranded (and therefore will not self anneal), and can be used for very sensitive detection of DNA or RNA.

Ribosome, A cellular particle which is involved in the translation of mRNAs to make proteins. Ribosomes are a complex consisting of ribosomal RNAs (rRNA) and several proteins.

RNA, A chemical found in the nucleus and cytoplasm of cells; it plays an important role in protein synthesis and other chemical activities of the cells from which all blood cells develop.

RNAi, 'RNA interference' (a.k.a. 'RNA silencing') is the mechanism by which small double-stranded RNAs can interfere with expression of any mRNA having a similar sequence. Those small RNAs are known as 'siRNA', for short interfering RNAs. The mode of action for siRNA appears to be via dissociation of its strands, hybridization to the target RNA, extension of those fragments by an RNA-dependent RNA polymerase, then fragmentation of the target. Importantly, the remnants of the target molecule appears to then act as an siRNA itself; thus the effect of a small amount of starting siRNA is effectively amplified and can have long-lasting effects on the recipient cell.

RNase protection assay, This is a sensitive method to determine (1) the amount of a specific mRNA present in a complex mixture of mRNA and/or (2) the sizes of exons which comprise the mRNA of interest. A radioactive DNA or RNA probe (in excess) is allowed to hybridize with a sample of mRNA (for example, total mRNA isolated from tissue), after which the mixture is digested with single-strand specific nuclease. Only the probe which is hybridized to the specific mRNA will escape the nuclease treatment, and can be detected on a gel. The amount of radioactivity which was protected from nuclease is proportional to the amount of mRNA to which it hybridized. If the probe included both intron and exons, only the exons will be protected from nuclease and their sizes can be ascertained on the gel.

RNase, Ribonuclease; an enzyme which degrades RNA. It is ubiquitous in living organisms and is exceptionally stable. The prevention of RNase activity is the primary problem in handling RNA.

rRNA, "ribosomal RNA"; any of several RNAs which become part of the ribosome, and thus are involved in translating mRNA and synthesizing proteins. They are the most abundant RNA in the cell (on a mass basis).

Runoff, n: Surface water effluent (usually from precipitation but may be from human activities such as irrigation) that moves too quickly to be absorbed into the ground. It flows down contour gradients to enter stream and river systems, carrying with it anything light enough to be borne in the volume of water, which may be light after a small rain or tremendous in the wake of a storm, when even large boulders and trees get swept up in the runoff. When runoff travels over *deforested* or unplanted agricultural lands, it carries away large quantities of *topsoil* . Runoff from agricultural areas often carries heavy doses of *biocides* , fertilizers, and other nutrients, which can lead to *eutrophication* when introduced into aquatic systems.

S1 end mapping, A technique to determine where the end of an RNA transcript lies with respect to its template DNA (the gene). Can't be described in a short paragraph. See "RNAse Protection assay" for a closely related technique.

S1 nuclease, An enzyme which digests only single-stranded nucleic acids.

Salinization, n: Accumulation of salts in soil that can eventually make the soil unable to support plant growth.

Screening, To screen a library (see "Library") is to select and isolate individual clones out of the mixture of clones. For example, if you needed a cDNA clone of the pituitary glycoprotein hormone alpha subunit, you would need to make (or buy) a pituitary cDNA library, then screen that library in order to detect and isolate those few bacteria carrying alpha subunit cDNA. There are two methods of screening which are particularly worth describing: screening by hybridization, and screening by antibody. Screening by hybridization involves spreading the mixture of bacteria out on a dozen or so agar plates to grow several ten thousand isolated colonies. Membranes are laid onto each plate, and some of the bacteria from each colony stick, producing replicas of each colony in their original growth position. The membranes are lifted and the adherent bacteria are lysed, then hybridized to a radioactive piece of alpha DNA (the source of which is a story in itself - see "Probe"). When X-ray film is laid on the filter, only colonies carrying alpha sequences will "light up". Their position on the membranes show where they grew on the original plates, so you now can go back to the original plate (where the remnants of the colonies are still alive), pick the colony off the plate and grow it up. You now have an unlimited source of alpha cDNA.

Second law of thermodynamics, n: In any conversion of heat energy to useful work, some of the initial energy input is always degraded to lower quality, more dispersed, less useful energy — usually low-temperature heat that flows into the environment; every energy system has "leaks" and looses energy or heat to attenuation.

Sequence, As a noun, the sequence of a DNA is a buzz word for the structure of a DNA molecule, in terms of the sequence of bases it contains. As a verb, "to sequence" is to determine the structure of a piece of DNA; i.e. the sequence of nucleotides it contains.

Shotgun cloning, The practice of randomly clipping a larger DNA fragment into various smaller pieces, cloning everything, and then studying the resulting individual clones to figure out what happened. For example, if one was studying a 50 kb gene, it "may" be a bit difficult to figure out the restriction map. By randomly breaking it into smaller fragments and mapping those, a master restriction map could be deduced. See also Shotgun sequencing.

Shotgun sequencing, A way of determining the sequence of a large DNA fragment which requires little brainpower but lots of late nights. The large fragment is shotgun cloned (see above), and then each of the resulting smaller clones ("subclones") is sequenced. By finding out where the subclones overlap, the sequence of the larger piece becomes apparent. Note that some of the regions will get sequenced several times just by chance.

siRNA, Small Inhibitory RNA; a.k.a. 'RNAi'. See 'RNAi'.

Slot blot, Similar to a dot blot, but the analyte is put onto the membrane using a slot-shaped template. The template produces a consistently shaped spot, thus decreasing errors and improving the accuracy of the analysis. See Dot blot.

SNP, Single Nucleotide Polymorphism (SNP) - a position in a genomic DNA sequence that varies from one individual to another. It is thought that the primary source of genetic difference between any two humans is due to the presence of single nucleotide polymorphisms in their DNA. Furthermore, these SNPs can be extremely useful in genetic mapping (see 'Genetic Mapping') to follow inheritance of specific segments of DNA in a lineage. SNP-typing is the process of determining the exact nucleotide at positions known to be polymorphic.

snRNA, Small nuclear RNA; forms complexes with proteins to form snRNPs; involved in RNA splicing, polyadenylation reactions, other unknown functions (probably).

snRNP, "snerps", Small Nuclear RiboNucleoProtein particles, which are complexes between small nuclear RNAs and proteins, and which are involved in RNA splicing and polyadenylation reactions.

Soil Erosion, n: The loss of topsoil through silt-laden *runoff* , strong winds, or other forces that transport soil away from its natural location.

Solution hybridization, A method closely related to RNase protection (see "RNase protection assay"). Solution hybridization is designed to measure the levels of a specific mRNA species in a complex population of RNA. An excess of radioactive probe is allowed to hybridize to the RNA, then single-strand specific nuclease is used to destroy the remaining unhybridized probe and RNA. The "protected" probe is separated from the degraded fragments, and the amount of radioactivity in it is proportional to the amount of mRNA in the sample which was capable of hybridization. This can be a very sensitive detection method.

Southern blot, A technique for analyzing mixtures of DNA, whereby the presence and rough size of one particular fragment of DNA can be ascertained. See "Blotting". Named for its inventor, Dr E. M. Southern.

Specialist species, n: Species with a narrow ecological niche. They may be able to live in only one type of habitat, tolerate only a narrow range of climatic or other environmental conditions, or they may use only one or a few types of food.

Speciation, n: Formation of two species from one species as a result of divergent natural selection in response to changes in environmental conditions; usually takes thousands or tens of thousands of years.

Species, n: The boundaries of this taxonomic level (the most precise in the hierarchical system of *binomial nomenclature*) are hotly debated among scientists and there is little real consensus about where to draw the lines between species, subspecies, morphs, races, variants, etc. In general, a species is a group of organisms that resemble one another in appearance, general behavior, ecological niche, chemical makeup and processes, and genetic structure. Organisms that reproduce sexually are classified as members of the same species only if they can actually or potentially interbreed with one another and produce fertile offspring. It should be noted that some (though quite few) taxonomists believe the species level of classification is frequently invalid and these scientists only recognize classifications down to the level of genus (again, these taxonomists represent a very small minority view).

SSR, Simple Sequence Repeat. See 'Microsatellite'.

Stable transfection, A form of transfection experiment designed to produce permanent lines of cultured cells with a new gene inserted into their genome. Usually this is done by linking the desired gene with a "selectable" gene, i.e. a gene which confers resistance to a toxin (like G418, aka Geneticin). Upon putting the toxin into the culture medium, only those cells which incorporate the resistance gene will survive, and essentially all of those will also have incorporated the experimenter's gene.

Sticky ends, After digestion of a DNA with certain restriction enzymes, the ends left have one strand overhanging the other to form a short (typically 4 nt) single-stranded segment. This overhang will easily re-attach to other ends like it, and are thus known as "sticky ends". For example, the enzyme BamHI recognizes the sequence GGATCC, and clips after the first G in each strand:

Stringency, A term used to describe the conditions of hybridization. By varying the conditions (especially salt concentration and temperature) a given probe sequence may be allowed to hybridize only with its exact complement (high stringency), or with any somewhat related sequences (relaxed or low stringency). Increasing the temperature or decreasing the salt concentration will tend to increase the selectivity of a hybridization reaction, and thus will raise the stringency.

Sub-cloning, If you have a cloned piece of DNA (say, inserted into a plasmid) and you need unlimited copies of only a part of it, you might "sub-clone" it. This involves starting with several million copies of the original plasmid, cutting with restriction enzymes, and purifying the desired fragment out of the mixture. That fragment can then be inserted into a new plasmid for replication. It has now been subcloned.

Sulfur cycle, n: Cyclic movement of sulfur in different chemical forms, from the environment to organisms and then back to the environment.

Sustainability, n: Ability of a system to survive for some specified (finite) time.

Sustainable agriculture, n: Method of growing crops and raising livestock based on organic fertilizers, soil conservation, water conservation, biological control of pests, and minimal use of non-renewable fossil-fuel energy.

Sustainable development, n: Forms of economic development and activities that do not deplete or degrade the natural resources upon which present and future economic growth and life depend.

Sustainable living, n: Taking no more potentially renewable resources from the natural world than can be replenished naturally and not overloading the capacity of the environment to cleanse and renew itself by natural processes.

Sustainable society, n: A society that manages its economy and population size without doing irreparable environmental harm by overloading the planet's ability to absorb environmental insults, replenish its resources, and sustain human and other forms of life over a specified period-usually hundreds to thousands of years. During this period it satisfies the needs of its people without depleting earth capital and thereby jeopardizing the prospects of current and future generations of humans and other species.

Sustainable system, n: A system that survives and functions over some specified (finite) time; a system that attains its full expected lifetime.

Sustainable yield (sustained yield), n: Highest rate at which a potentially renewable resource can be used without reducing its available supply throughout the world or in a particular area.

Symbiosis, n: Literally means "living together" in Latin. Any intimate relationship or association between members of two or more species. The members of the relationship are symbionts. Obligate symbionts rely so heavily on the relationship that they cannot feed, reproduce, or perform some other crucial life function in the absence of their symbiotic partner(s). There are three main categories of symbiosis: *commensalism* , *mutualism* , and *parasitism* , with some degree of blending at the edges of these definitions in many cases.

Symbiotic, adj: Refers to a component or member of a system of symbiosis. "These organisms have a symbiotic relationship."

Taq polymerase, A DNA polymerase isolated from the bacterium *Thermophilis aquaticus* and which is very stable to high temperatures. It is used in PCR procedures and high temperature sequencing.

TATA box, A sequence found in the promoter (part of the 5' flanking region) of many genes. Deletion of this site (the binding site of transcription factor TFIID) causes a marked reduction in transcription, and gives rise to heterogeneous transcription initiation sites.

Taxonomy, n: The classification of living organisms according to the heirarchy of relationships.

Telomere, The ends of chromosomes. These specialized structures are involved in the replication and stability of linear DNA molecules.

Tet resistance, See "Antibiotic resistance".

The overhangs thus produced can still hybridize ("anneal") with each other, even if they came from different parent DNA molecules, and the enzyme ligase will then covalently link the strands. Sticky ends therefore facilitate the ligation of diverse segments of DNA, and allow the formation of novel DNA constructs.

The RNAi effect has been exploited in numerous research programs to deplete the call of specific messages, thus examining the role of those messages by their absence.

Threatened species, n: Wild species that is still abundant in its natural range but is likely to become endangered because of a decline in numbers.

Tissue-specific expression, Gene function which is restricted to a particular tissue or cell type. For example, the glycoprotein hormone alpha subunit is produced only in certain cell types of the anterior pituitary and placenta, not in lungs or skin; thus expression of the glycoprotein hormone alpha-chain gene is said to be tissue-specific. Tissue specific expression is usually the result of an enhancer which is activated only in the proper cell type.

Tm, The melting point for a double-stranded nucleic acid. Technically, this is defined as the temperature at which 50% of the strands are in double-stranded form and 50% are single-stranded, i.e. midway in the melting curve. A primer has a specific Tm because it is assumed that it will find an opposite strand of appropriate character.

Total fertility rate (TFR), n: Estimate of the average number of children that will be born alive to a woman during her lifetime if she passes through all her childbearing years (ages 15-44) conforming to age-specific fertility rates of a given year. In simpler terms, it is an estimate of the average number of children a woman will have during her childbearing years.

Transcription factor, A protein which is involved in the transcription of genes. These usually bind to DNA as part of their function (but not necessarily). A transcription factor may be general (i.e. acting on many or all genes in all tissues), or tissue-specific (i.e. present only in a particular cell type, and activating the genes restricted to that cell type). Its activity may be constitutive, or may depend on the presence of some stimulus; for example, the glucocorticoid receptor is a transcription factor which is active only when glucocorticoids are present.

Transcription, The process of copying DNA to produce an RNA transcript. This is the first step in the expression of any gene. The resulting RNA, if it codes for a protein, will be spliced, polyadenylated, transported to the cytoplasm, and by the process of translation will produce the desired protein molecule.

Transcription, The process of copying information from DNA into new strands of messenger RNA (mRNA).

Transfection, A method by which experimental DNA may be put into a cultured mammalian cell. Such experiments are usually performed using cloned DNA containing coding sequences and control regions (promoters, etc) in order to test whether the DNA will be expressed. Since the cloned DNA may have been extensively modified (for example, protein binding sites on the promoter may have been altered or removed), this procedure is often used to test whether a particular modification affects the function of a gene.

Transformation (with respect to bacteria), The process by which a bacteria acquires a plasmid and becomes antibiotic resistant. This term most commonly refers to a bench procedure performed by the investigator which introduces experimental plasmids into bacteria.

Transformation (with respect to cultured cells), A change in cell morphology and behavior which is generally related to carcinogenesis. Transformed cells tend to exhibit characteristics known collectively as the "transformed phenotype" (rounded cell bodies, reduced attachment dependence, increased growth rate, loss of contact inhibition, etc). There are different "degrees" of transformation, and cells may exhibit only a subset of these characteristics. Not well understood, the process of transformation is the subject of intense research.

Transgenic mouse, A mouse which carries experimentally introduced DNA. The procedure by which one makes a transgenic mouse involves the injection of DNA into a fertilized embryo at the pro-nuclear stage. The DNA is generally cloned, and may be experimentally altered. It will become incorporated into the genome of the embryo. That embryo is implanted into a foster mother, who gives birth to an animal carrying the new gene. Various experiments are then carried out to test the functionality of the inserted DNA.

Transient transfection, When DNA is transfected into cultured cells, it is able to stay in those cells for about 2-3 days, but then will be lost (unless steps are taken to ensure that it is retained - see Stable transfection). During those 2-3 days, the DNA is functional, and any functional genes it contains will be expressed. Investigators take advantage of this transient expression period to test gene function.

Translation, The process of decoding a strand of mRNA, thereby producing a protein based on the code. This process requires ribosomes (which are composed of rRNA along with various proteins) to perform the synthesis, and tRNA to bring in the amino acids. Sometimes, however, people speak of "translating" the DNA or RNA when they are merely reading the nucleotide sequence and predicting from it the sequence of the encoded protein. This might be more accurately termed "conceptual translation".

Translation, The process of turning instructions from mRNA, base by base, into chains of amino acids that then fold into proteins.

Transmissible disease, n: A disease that is caused by living organisms (such as bacteria, viruses, and parasitic worms) and that can spread from one person to another by air, water, food, body fluids (or in some cases by insects or other organisms).

tRNA, "transfer RNA"; one of a class of rather small RNAs used by the cell to carry amino acids to the enzyme complex (the ribosome) which builds proteins, using an mRNA as a guide. Fairly abundant.

Tumor suppressor, A gene that inhibits progression towards neoplastic transformation. The best-known examples of tumor suppressors are the proteins **p53** and **Rb**.

Umbilical cord, The structure that connects the placenta and the embryo; contains the umbilical arteries and the umbilical vein.

Unicellular, Single-celled.

Uniformitarianism, The idea that geological processes have remained uniform over time and that slight changes over long periods can have large-scale consequences; proposed by James Hutton in 1795 and reÞned by Charles Lyell during the 1800s. The principle on which modern geology was founded: processes operating today on the earth operated in much the same way in the geologic past. Sometimes expressed as "the present is the key to the past".

Uninucleate, Term applied to cells having only a single nucleus.

Unsaturated fat, A triglyceride with double coavent bonds between some carbon atoms.

Upstream activator sequence, A binding site for transcription factors, generally part of a promoter region. A UAS may be found upstream of the TATA sequence (if there is one), and its function is (like an enhancer) to increase transcription. Unlike an enhancer, it can not be positioned just anywhere or in any orientation.

Upstream/Downstream, In an RNA, anything towards the 5' end of a reference point is "upstream" of that point. This orientation reflects the direction of both the synthesis of mRNA, and its translation - from the 5' end to the 3' end. In DNA, the situation is a bit more complicated. In the vicinity of a gene (or in a cDNA), the DNA has two strands, but one strand is virtually a duplicate of the RNA, so it's 5' and 3' ends determine upstream and downstream, respectively. NOTE that in genomic DNA, two adjacent genes may be on different strands and thus oriented in opposite directions. Upstream or downstream is only used on conjunction with a given gene.

Uracil, The pyrimidine that replaces thymine in RNA molecules and nucleotides.

Urban heat island, n: Buildup of heat in the atmosphere above an urban area. This is produced by the large concentration of cars, buildings, factories, and other heat-producing activities.

Ureter, A muscular tube that transports urine by peristaltic contractions from the kidney to the bladder.

Urethra, A narrow tube that transports urine from the bladder to the outside of the body. In males, it also conducts sperm and semen to the outside.

Urine, Fluid containing various wastes that is produced in the kidney and excreted from the bladder.

Uterus, The organ that houses and nourishes the developing embryo and fetus. The womb. Female reproductive organ in which the fertilized egg implants.

Vaccination, The process of protecting against infectious disease by introducing into the body a vaccine that stimulates a primary immune response and the production of memory cells against the disease-causing agent.

Vaccine, A preparation containing dead or weakened pathogens that when injected into the body elicit an immune response.

Vacuoles, Membrane-bound þuid-Þlled spaces in plant and animal cells that remove waste products and store ingested food.

Vagina, The tubular organ that is the site of sperm deposition and also serves as the birth canal.

Vas deferens, The duct that carries sperm from the epididymis to the ejaculatory duct and urethra. The tube connecting the testes with the urethra.

Vascular bundle, Groups of xylem, phloem and cambium cells in stems of plants descended from the procambium embryonic tissue layer.

Vascular cambium, A layer of lateral meristematic tissue between the xylem and phloem in the stems of woody plants. Lateral meristem tissue in plants that produces secondary growth.

Vascular cylinder, A central column formed by the vascular tissue of a plant root; surrounded by parenchymal ground tissue.

Vascular parenchyma, Specialized parenchyma cells in the phloem of plants.

Vascular plants, Group of plants having lignified conducting tissue (xylem vessels or tracheids).

Vascular system, Specialized tissues for transporting þuids and nutrients in plants; also plays a role in supporting the plant; one of the four main tissue systems in plants.

Vector, The DNA "vehicle" used to carry experimental DNA and to clone it. The vector provides all sequences essential for replicating the test DNA. Typical vectors include plasmids, cosmids, phages and YACs.

Ventricle, The chamber of the heart that pumps the blood into the blood vessels that carry it away from the heart. The lower chamber of the heart through which blood leaves the heart.

Venules, The smallest veins. Blood þows into them from the capillary beds. Small veins that connect a vein with capillaries.

Vernalization, ArtiÞcial exposure of seeds or seedlings to cold to enable the plant to þower.

Vertebrae, The segments of the spinal column; separated by disks made of connective tissue (sing.: vertebra).

Vertebrate, Any animal having a segmented vertebral column; members of the subphylum Vertebrata; include reptiles, Þshes, mammals, and birds.

Vesicles, Small membrane-bound spaces in most plant and animal cells that transport macromolecules into and out of the cell and carry materials between organelles in the cell.

Vessel elements, Short, wide cells arranged end to end, forming a system of tubes in the xylem that moves water and solutes from the roots to the rest of the plant. Large diameter cells of the xylem that are extremely specialized and efficient at conduction. An evolutionary advance over tracheids. Most angiosperms have vessels.

Vestigial structures, Nonfunctional remains of organs that were functional in ancestral species and may still be functional in related species; e.g., the dewclaws of dogs.

Villi, Finger-like projections of the lining of the small intestine that increase the surface area available for absorption. Also, projections of the chorion that extend into cavities Þlled with maternal blood and allow the exchange of nutrients between the maternal and embryonic circulations. Projections of the inner layer of the small intestine that increase the surface area for absorbtion of food.

Viroids, Infective forms of nucleic acid without a protective coat of protein; unencapsulated single-stranded RNA molecules. Naked RNA, possibly of degenerated virus, that infects plants.

Virus, Infectious chemical agent composed of a nucleic acid (DNA or RNA) inside a protein coat.

Vitamins, A diverse group of organic molecules that are required for metabolic reactions and generally cannot be synthesized in the body.

Vulva, A collective term for the external genitals in women.

Western blot, A technique for analyzing mixtures of proteins to show the presence, size and abundance of one particular type of protein. Similar to Southern or Northern blotting (see "Blotting"), except that (1) a protein mixture is electrophoresed in an acrylamide gel, and (2) the "probe" is an antibody which recognizes the protein of interest, followed by a radioactive secondary probe (such as 125I-protein A).

White blood cell, Component of the blood that functions in the immune system. Also known as a leukocyte.

Xerophytic leaves, The leaves of plants that grow under arid conditions with low levels of soil and water. Usually characterized by water-conserving features such as thick cuticle and sunken stomatal pits.

X-ray diffraction, Technique utilized to study atomic structure of crystalline substances by noting the patterns produced by x-rays shot through the crystal.

Xylem, Tissue in the vascular system of plants that moves water and dissolved nutrients from the roots to the leaves; composed of various cell types including tracheids and vessel elements. Plant tissue type that conducts water and nutrients from the roots to the leaves.

YAC, Yeast artificial chromosome. This is a method for cloning very large fragments of DNA. Genomic DNA in fragments of 200-500 kb are linked to sequences which allow them to propagate in yeast as a mini-chromosome (including telomeres, a centromere and an ARS - an autonomous replication sequence). This technique is used to clone large genes and intergenic regions, and for chromosome walking.

Z lines, Dense areas in myoÞbrils that mark the beginning of the sarcomeres. The actin Þlaments of the sarcomeres are anchored in the Z lines.

Zero population growth (ZPG), n: State in which the birth rate (plus immigration) equals the death rate (plus emigration) so that the population of a geographical area is no longer increasing.

Zinc finger, A protein structural motif common in DNA binding proteins. Four Cys residues are found for each "finger" and one finger can bind a molecule of zinc. A typical configuration is: CysXxxXxxCys—(intervening 12 or so aa's)—CysXxxXxxCys.

Zone of physiological stress The area in a population's geographic range where members of population are rare due to physical and biological limiting factors.

Zoonosis, n: A disease of animals, such as rabies or psittacosis, that can be transmitted to humans. **Zoonotic**, adj.

Zygospore, In fungi, a structure that forms from the diploid zygote created by the fusion of haploid hyphae of different mating types. After a period of dormancy, the zygospore forms sporangia, where meiosis occurs and spores form.

Zygote, A fertilized egg. A diploid cell resulting from fertilization of an egg by a sperm cell.

Bibliography

Ahlstedt, S.A., and Bakaletz, S. (2005). *Assessment of freshwater mussel populations in Clear Creek, Tennessee (Emory-Obed River System) following an oil well fire and resulting spill*. Prepared for the National Park Service, Atlanta, Georgia.

Ahlstedt, S.A., Connell, J.F., Bakaletz, S., and Fagg, M.T. (2001). *Freshwater mussels of the National Park Service's Obed Wild and Scenic River, Tennessee*. Prepared for the National Park Service, Oneida, Tennessee.

Alberts, Bruce; Johnson, A, Lewis, J, Raff, M, Roberts, K & Walter, P (2002). *Molecular Biology of the Cell*, 4th edition, Garland.

Alcorta, Candace S. and Richard Sosis (2005). Ritual, emotion, and sacred symbols: The evolution of religion as an adaptive complex. *Human Nature* 16(4): 323-359.

Aldenderfer, Mark (2006). Costly signaling, the sexual division of labor, and animal domestication in the Andean highlands. In *Behavioral Ecology and the Transition to Agriculture*, D.J. Kennett and B. Winterhalder, eds., pp. 167-196. Berkeley: University of California Press.

Allal, N., R. Sear, A.M. Prentice, and R. Mace (2004). An evolutionary model of stature, age at first birth and reproductive success in Gambian women. *Proceedings of the Royal Society of London Series B, Biological Sciences* 271(1538): 465-470.

Alvard, M., and L. Kuznar (2002). The transition from hunting to animal husbandry: Prey choice when harvests are deferred. *American Anthropologist* 103(2): 295-311.

Alvard, Michael S., and Allen Gillespie (2004). Good Lamalera whale hunters accrue reproductive benefits. *Research in Economic Anthropology* 23: 225-247.

Anderson, Kermyt G., and Bobbi S. Low (2003). Nonmarital first births and women's life histories. In *The Biodemography of Human Reproduction and Fertility*, J. Rodgers and H.P. Kohler, eds., pp. 57-86. Boston: Kluwer Academic Publishers.

Anderson, Kermyt G., Hillard Kaplan, and Jane B. Lancaster (2006). Demographic correlates of paternity confidence and pregnancy outcomes among Albuquerque men. *American Journal of Physical Anthropology* 131(4): 560-571.

Anderson, Kermyt G., Hillard Kaplan, and Jane B. Lancaster (2007). Confidence of paternity, divorce, and investment in children by Albuquerque men. *Evolution and Human Behavior* 28(1): 1-10.

Apicella, C.L., and Marlowe, F.W (2004). Perceived mate fidelity and paternal resemblance predict men's investment in children. *Evolution and Human Behavior* 25: 371-378.

Apicella, C.L., and Marlowe, F.W (2007). Men's reproductive decisions: Mating, parenting and self-perceived mate value. *Human Nature* 18:22-34.

Beck, W.S., Liem, K.P., Simpson, G.G. (1991) Life 3rd Edition. New York: Harper Collins.

Begon, Michael; Townsend, CR & Harper, JL (2005). *Ecology: From Individuals to Ecosystems*, 4th edition, Blackwell Publishing Limited.

Beise, Jan, and Eckart Voland (2002). Effect of producing sons on maternal longevity in premodern populations. *Science* 298: 317a.

Benyshek, Daniel C., and James T. Watson (2006). Exploring the thrifty genotype's food-shortage assumptions: A cross-cultural comparison of ethnographic accounts of food security among foraging and agricultural societies. *American Journal of Physical Anthropology* 131(1): 120-126.

Bereczkei, Tamas, Adam Hofer, and Zsuzsanna Ivan (2000). Low birth weight, maternal birth-spacing decisions, and future reproduction: A cost-benefit analysis. *Human Nature* 11: 183-205.

Bird Douglas W., Rebecca Bliege Bird, and C. H. Parker (2004). Women who hunt with fire. *Australian Aboriginal Studies* 1: 90-96.

Bird, D., and R. Bliege Bird (2003). Mardu children's hunting strategies in the Western Desert, Australia. In *Hunter-gatherer Children*, B. Hewlett ed., Senri Ethnological Studies: papers from the 9th International Conference on Hunting and Gathering Societies.

Bird, D.W (2002). Human foraging strategies: Human diet and food practices. In *Encyclopedia of Evolution*, M. Pagel, ed. New York: Oxford University Press.

Bird, D.W., and R. Bliege Bird (2002). Children on the reef: Slow learning or strategic foraging? *Human Nature* 13: 269-298.

Bird, D.W., J.L. Richardson, P.M. Veth, and A.J. Barham (2002). Explaining shellfish variability in middens on the Meriam Islands, Torres Strait Australia. *Journal of Archaeological Science* 29: 457-469.

Bird, Douglas W. and Rebecca Bliege Bird (2004). Evolutionary and ecological understandings of the economics of desert societies: Comparing the Great Basin USA and the Australian Deserts. In *Desert Peoples: Archaeological Perspectives*, P.M. Veth, M. Smith, and P. Hiscock (eds), London: Blackwell Scientific.

Bird, Douglas W., and Rebecca Bliege Bird (2005). Martu children's hunting strategies in the Western Desert, Australia. In *Hunter-Gatherer Childhoods: Evolutionary, Developmental, and Cultural Perspectives*, B.S. Hewlett and M.E. Lamb, eds. pp. 129-146. New Brunswick: AldineTransaction.

Bird, Douglas W., Rebecca Bliege Bird, and J. L. Richardson (2004). Meriam ethnoarchaeology: shellfishing and shellmiddens. *Memoirs of the Queensland Museum, Cultural Heritage Series* 3(1): 183-197.

Bird, Douglas, Rebecca Bliege Bird, and C. Parker (2005). Aboriginal burning regimes and hunting strategies in Australia's Western Desert. *Human Ecology* 33(4): 443-464.

Bishai, David, Nan Astone, Laura Argys, Robert Gutendorf, and Chris Filidoro (2006). A national sample of US paternity tests: Do demographics predict test outcomes? *Transfusion* 46(5): 849-863.

Blatt, Robin J.R. (1993) "The Human Genome Project" in association with MCET The Mass LearnPike, Cambridge, Massachusetts.

Bliege Bird, R., D.W. Bird, Geoff Kushnick, and E.A. Smith (2002). Risk and reciprocity in Meriam food sharing. *Evolution and Human Behavior* 23: 297-321.

Bliege Bird, R., E.A. Smith, and D.W. Bird (2002). The hunting handicap: Costly signaling in male foraging strategies. *Behavioral Ecology and Sociobiology* 50: 9-19.

Borgerhoff Mulder, M (2000). Optimizing offspring: The quality-quantity tradeoff in agropastoral Kipsigis. *Evolution and Human Behavior* 21(6) 390-410.

Borgerhoff Mulder, M (2002). Using phylogenetically controlled comparisons in anthropology: More questions than answers. *Evolutionary Anthropology* 10: 99-111.

Bowles, S., and D.Posel (2005). Genetic relatedness predicts South African migrant workers' remittances to their families. *Nature* 434(7031): 380-383.

Bribiescas, Richard G (2006). On the evolution, life history, and proximate mechanisms of human male reproductive senescence. *Evolutionary Anthropology* 15(4): 132-141.

Bright, J. (2002). Institutional constraints on economic and social fluidity in farmer-forager systems: Bioarchaeology and the sexual division of labor in prehistoric Utah. *Utah Archaeology* 15: 67-84.

Brock, T. and Madigan, M (1992) "Gene manipulation and genetic engineering," In: Biology of Microorganisms 6th Edition. Englewood Cliffs, NJ: Prentice Hall, pp. 280-305.

Burnham, T.C., Flynn Chapman, J.C., Gray, P.B., McIntyre, M., Lipson, S.F. and

Ellison, P.T (2003). Men in committed, romantic relationships have lower testosterone than single men. *Hormones and Behavior* 44: 119-122.

Caldwell, John C. (2005). On net intergenerational wealth flows: An update. *Population and Development Review* 31(4): 721-740.

Calfee, John E. and Elizabeth DuPre (2006). *"The Emerging Market Dynamic of Targeted Therapeutics," Health Affairs*. 25(5): 1302-1308.

Camerer, C., and E. Fehr (2004). Measuring social norms and preferences using experimental games anthropologists. In *Foundations of Human Sociality: Ethnography and Experiments in 15 Small-scale Societies*. J. Henrich, R. Boyd, S. Bowles, H. Gintis, E. Fehr, and Colin Camerer (eds.). Oxford University Press.

Campbell, Benjamin (2006). Adrenarche and the evolution of human life history. *American Journal of Human Biology* 18(5): 569-589.

Campbell, Neil (2004). *Biology*, 7th edition, Benjamin-Cummings Publishing Company.

Cannon, M. (2003). A model of central place forager prey choice and an application to faunal remains from the Mimbres Valley, New Mexico. *Journal of Anthropological Archaeology* 22: 1-25.

Chisholm, James S., Julie A. Quinlivan, Rodney W. Petersen, and David A. Coall (2005). Early stress predicts age at menarche and first birth, adult attachment and expectedlifespan. *Human Nature* 16(3): 233-265.

Chrastil ER, Getz WM, Euler HA, Starks PT (2006). Paternity uncertainty overrides sex chromosome selection for preferential grandparenting. *Evolution and Human Behavior* 27(3): 206-223.

Clarke, Alice L., and Bobbi S. Low (2002). Testing evolutionary hypotheses with demographic data. *Population and Development Review* 27(4): 663-660.

Collardad, Mark, Stephen J. Shennan, and Jamshid J. Tehrani (2006). Branching, blending, and the evolution of cultural similarities and differences among human populations. *Evolution and Human Behavior* 27(3): 169-184.

Cook, R.B., Suter, II, G.W., and Sain, E.R. (1998). Ecological risk assessment in a large river-reservoir: 1. Introduction and background [Electronic version]. *Environmental Toxicology and Chemistry*, 18(4), 581-588.

Cook, S.B. (2006). *Quantification of injury to aquatic resources resulting from the Pryor oil spill and fire, Obed Wild and Scenic River, Task 1* (pdf file). Final Draft Report.

Cook, S.B., Russ, W.T., and Goodfred, D.W. (2005, April). *Emory River Watershed biological assessment* (pdf file). Quarterly Report submitted to the U.S. Fish and Wildlife Service and Tennessee Wildlife Resources Agency.

Cook, S.B., Russ, W.T., and Goodfred, D.W. (2005, February). Emory River Watershed biological assessment (pdf file). Quarterly Report submitted to the U.S. Fish and Wildlife Service and Tennessee Wildlife Resources Agency.

Cook, S.B., Russ, W.T., and Goodfred, D.W. (2006, September). *Biological survey of the Emory River Watershed, Tennessee* (pdf file). Final Report submitted to the U.S. Fish and Wildlife Service and Tennessee Wildlife Resources Agency.

Cordain, L., S.B. Eaton, J.B. Miller, N. Mann, and K. Hill (2002). The paradoxical nature of hunter-gatherer diets: Meat based yet non-atherogenic. *European J. of Clinical Nutrition* 56 suppl. 1: 542-552.

Cristianini, N. and Hahn, M. (2006). *Introduction to Computational Genomics*, Cambridge University Press.

Crognier, E. (2003). Reproductive success: Which meaning? *American Journal of Human Biology*: 15(3): 352-360.

Crognier, E., A. Baali, M.-K. Hilali, M. Villena, E. Vargas (2006). Preference for sons and sex ratio in two non-western societies. *American Journal of Human Biology* 18: 325-334.

Crognier, E., M. Villena, and E. Vargas (2002). Helping patterns and reproductive success in Aymara communities. *American Journal of Human Biology* 14(3): 372-379.

Crognier, E., M. Villena, and E. Vargas (2002). Reproduction in high altitude Aymara: Physiological stress and fertility planning? *Journal of Biosocial Science* 34(4): 463-473.

Cronk, Lee (2002). Anthropology and the evolutionary study of human behavior. *Research in Biopolitics* 8: 1-30.

Cronk, Lee (2004). Continuity, displaced reference, and deception. *Behavioral and Brain Sciences* 27(4): 510-511.

Cronk, Lee (2004). *From Mukogodo to Maasai: Ethnicity and Cultural Change in Kenya*. Boulder, CO: Westview Press.

Cronk, Lee (2005). The application of animal signaling theory to human phenomena: Some thoughts and clarifications. *Social Science Information/Information sur les Sciences Sociales* 44(4): 603-620.

Cronk, Lee (2006). Behavioral ecology and the social sciences. In *Missing the Revolution: Evolutionary Psychology for Social Scientists*, edited by Jerome Barkow. Pp. 167-185. Oxford: Oxford University Press.

Cronk, Lee and Shannon Steadman (2002). Tourists as a common-pool resource: A study of dive shops on Utila, Honduras. In *Economic Development: An Anthropological Approach*, edited by Jeffrey H. Cohen and Norbert Dannhaeuser, pp. 51-68. Society for Economic Anthropology Monographs, Volume 19. Walnut Creek: Altamira Press.

Damage assessment study plan: Pryor Oil Well fire and spill, Obed Wild and Scenic River (pdf file). (2004, September 30). Department of the Interior, State of Tennessee, National Park Service, U.S. Fish and Wildlife Service.

Deady, Denis K., Miriam J. Law Smith, J. P. Kent, and R. I. M. Dunbar (2006). Is priesthood an adaptive strategy? Evidence from a historical Irish population. *Human Nature* 17(4): 393-404.

Denham, Tim, and Huw Barton (2006). The emergence of agriculture in New Guinea: A model of continuity from pre-existing foraging practices. In *Behavioral Ecology and the Transition to Agriculture*, D.J. Kennett and B. Winterhalder, eds., pp. 237-264. Berkeley: University of California Press.

Department of the Interior, Fish and Wildlife Service. (2003, June 3). *Federal Register* (pdf file). 68(106). Retrieved from http://www.fws.gov/policy/library/03-12944.html

Department of the Interior, Fish and Wildlife Service. (2002, September 27). *Federal Register* (pdf file). 67(188). Retrieved from http://www.fws.gov/policy/library/02fr61016.html

Department of the Interior, Fish and Wildlife Service. (2004, August 31). *Federal Register* (pdf file). 69(168). Retrieved from http://www.fws.gov/policy/library/04-19340.html

Department of the Interior, Fish and Wildlife Service. (2005, April 8). *Federal Register* (pdf file). 70(67). Retrieved from http://www.fws.gov/policy/library/05-7086.html

Dickins, T.E., Sear, R. & Wells, A.J (2007). Mind the gap(s)...in theory, method and data: Re-examining Kanazawa (2006). *British Journal of Health Psychology* 12: 167-178.

Dodds, John H. (1985) *Plant Genetic Engineering*. New York: Cambridge University Press.

Draper, Patricia, and Raymond Hames (2000). Birth order, sibling investment, and fertility among Ju/'hoansi (!Kung). *Human Nature* 11: 117-156.

Dufour, Darna L (2006). Biocultural approaches in human biology. *American Journal of Human Biology* 18(1): 1-9.

Ellison, Peter T (2005). Evolutionary perspectives on the fetal origins hypothesis. *American Journal of Human Biology* 17(1): 113-118.

Emory River Watershed Development Association and TVA. (1968). *Emory River Valley: summary of resources* .

Ensminger, Jean (2004). Market Integration and Fairness: Evidence from ultimatum, dictator, and public goods experiments in East Africa. In *Foundations of Human Sociality: Ethnography and Experiments in 15 Small-scale Societies*. J. Henrich, R. Boyd, S. Bowles, H. Gintis, E. Fehr, and Colin Camerer (eds.). Oxford University Press.

Evans, J., Pelkey, D., and Haskell, D. (2002). *An assessment of forest change on the Cumberland Plateau in southern Tennessee: Small area forestry demonstration project for the Southern Forest Resource Assessment.* Report on file with the U.S. Environmental Protection Agency and the U.S. Fish and Wildlife Service.

Fehr, Ernst, and Joseph Henrich (2003). Is strong reciprocity a maladaptation. In *Genetic and Culture Evolution of Cooperation*, edited by Peter Hammerstein. Cambridge, MA: MIT Press.

Fehr, Ernst, Urs Fischbacher, and Simon Gächter (2002). Strong reciprocity, human cooperation, and the enforcement of social norms. *Human Nature* 13: 1-26.

Fessler, Daniel M.T., C. David Navarrete, William Hopkins, and M. Kay Izard (2005). Examining the terminal investment hypothesis in humans and chimpanzees: Associations among maternal age, parity, and birth weight. *American Journal of Physical Anthropology* 127(1): 95-104.

Fortunato, Laura, Clare Holden, and Ruth Mace (2006). From bridewealth to dowry?: A Bayesian estimation of ancestral states of marriage transfers in Indo-European groups. *Human Nature* 17(4): 355-376.

Foster, Caroline (2000). The limits to low fertility: A biosocial approach. *Population and Development Review* 26: 209-234.

Fouts, Hillary N (2004). Images of weaning: Social and emotional contexts of weaning among farmers and foragers. *Ethnology* 43(1): 65-81.

Fouts, Hillary N., and Michael E. Lamb (2005). Weanling emotional patterns among the Bofi foragers of Central Africa: The role of maternal availability and sensitivity. In *Hunter-Gatherer Childhoods*, edited by Barry S. Hewlett and Michael E. Lamb (pp. 309-321). New York: Aldine De Gruyter.

Fouts, Hillary N., Barry S. Hewlett, and Michael E. Lamb (2002). Weaning and the nature of early childhood interactions among Bofi foragers in central Africa. *Human Nature* 12: 27-46.

Fouts, Hillary N., Barry S. Hewlett, and Michael E. Lamb (2005). Parent-offspring weaning conflicts among the Bofi farmers and foragers of central Africa. *Current Anthropology* 46(1): 29-50.

Fouts, Hillary N., Michael E. Lamb, and Barry S. Hewlett (2004). Infant crying in hunter-gatherer cultures (comment). *Behavioral and Brain Sciences* 27(4): 462-463.

Fouts, Hillary, N. Father involvement with Bofi forager children: The role of grandmothers, residence, and marriage.

Friedberg, E.C. (1985) *DNA Repair.* New York: WH Freeman and Company.

Fruton, JS (1999). *Proteins, Enzymes, Genes: The Interplay of Chemistry and Biology.* Yale University Press.

Gibson, Mhairi A., and Ruth Mace (2002). Labor-saving technology and fertility increase in rural Africa. *Current Anthropology* 43(4): 631-337.

Gibson, Mhairi A., and Ruth Mace (2005). Helpful grandmothers in rural Ethiopia: A study of the effect of kin on child survival and growth. *Evolution and Human Behavior* 26(6): 469-482.

Gil-White, Francisco (2004). Ultimatum game with an ethnicity manipulation: Results from Bulgan Cum, Mongolia. In *Foundations of Human Sociality: Ethnography and Experiments in 15 Small-scale Societies*. J. Henrich, R. Boyd, S. Bowles, H. Gintis, E. Fehr, and Colin Camerer (eds.). Oxford University Press.

Goodfred, D.W. (2005, November). Characterization of benthic macroinvertebrate communities within the Emory River Watershed, Tennessee . In *The American Fisheries Society Fifth Annual Student Colloquium abstracts book*. Retrieved December 5, 2006, from http://www.ecu.edu/org/afs/5th%20Annual%20AFS%20Student%20Colloquium%20-%20Abstract%20Book.doc

Goodfred, D.W. (2006). *Characterization of benthic macroinvertebrate communities within the Emory River Watershed, Tennessee* (pdf file). Master's Thesis. Tennessee Technological University: Cookeville.

Goodman, D.C. (1987) *From Farming to Biotechnology: A Theory of Agro-industrial Development*. Oxford: Blackwell.

Goss, S. (2006, July). The Emory River Watershed (including the Obed). *Tennessee Citizens for Wilderness Planning*, 268. Retrieved September 11, 2006, from http://www.tcwp.org/nl268.shtml

Gray, P.B (2004). Evolutionary and cross-cultural perspectives on gambling. *Journal of Gambling Studies* 20: 347-371.

Gray, P.B., Campbell, B.C., Marlowe, F., Lipson, S.F. and Ellison, P.T (2004). Social variables predict between- subject but not day-to-day variation in the testosterone of U.S. men. *Psychoneuroendocrinology* 29: 1153-1162.

Gray, P.B., Kahlenberg, S., Barrett, E., Lipson, S. and Ellison, P.T (2002). Marriage and fatherhood are associated with lower testosterone in males. *Evolution and Human Behavior* 23: 193-201.

Gray, P.B., Singh, A., Woodhouse, L., Storer, T.W., Casaburi, R., Sinha-Hikim, I. Dzekov, C., Dzekov J. and Bhasin, S (2005). Effects of testosterone on sexual function, mood and visuospatial cognition in older men. *Journal of Clinical Endocrinology and Metabolism* 90: 3838-3846.

Gray, Sandra, Mary Sundal, Brandi Wiebusch, Michael A. Little, Paul W. Leslie, and Ivy L. Pike (2003). Cattle raiding, cultural survival, and adaptability of East African pastoralists. *Current Anthropology* 44(S5): S3-S30.

Greaves, R. Fishing strategies and technology: Ethnoarchaeology of resource structure influences on behavioral ecology of Pumé foragers in Venezuela.

Green, James, D; Tsang, Lincoln; Cavagnaro, Joy ((2003)). "'Generic' or 'Follow-on' Biologics: Scientific Considerations and Safety Issues," *Expert Opin. Biol. Ther.*; 3(7):1019-1022.

Gremillion, Kris (2002). Foraging theory and hypothesis testing in archaeology: An exploration of methodological problems and solutions. *Journal of Anthropological Archaeology* 21: 142-164.

Gremillion, Kris (2006). Central place foraging and food production on the Cumberland Plateau, Eastern Kentucky. In *Behavioral Ecology and the Transition to Agriculture*, D.J. Kennett and B. Winterhalder, eds., pp. 41-62. Berkeley: University of California Press.

Gurven, M (2004). Does market exposure affect economic game behavior? The ultimatum game and the public goods game among the Tsimane' of Bolivia. In *Foundations of Human Sociality: Ethnography and Experiments in 15 Small-scale Societies*. J. Henrich, R. Boyd, S. Bowles, H. Gintis, E. Fehr, and Colin Camerer (eds.). Oxford University Press.

Gurven, M. Economic games among the Amazonian Tsimane: Exploring the roles of market access, costs of giving, and cooperation on pro-social game behavior.

Hagen, E.H., and G.A. Bryant (2003). Music and dance as a coalition signaling system. *Human Nature* 14: 21-51.

Hagen, E.H., and P. Hammerstein (2005). Evolutionary biology and the strategic view of ontogeny: Genetic strategies provide robustness and flexibility in the life course. *Research in Human Development* 2: 87-101.

Hagen, E.H., H.C. Barrett, and M.E. Price (2006). Do human parents face a quantity-quality tradeoff? Evidence for a quantity-quality tradeoff. *American Journal of Physical Anthropology* 130: 405-418.

Hagen, E.H., R.B. Hames, N.M. Craig, M.T. Lauer, and M.E. Price (2002). Parental investment and child health in a Yanomamö village suffering short-term food stress. *Journal of Biosocial Science* 33: 503-528.

Haig, David (2003). On intrapersonal reciprocity. *Evolution and Human Behavior* 24(6): 418-425.

Hall, Roberta (2004). An energetics-based approach to understanding the menstrual cycle and menopause. *Human Nature* 15(1): 83-99.

Hames, Raymond (2000). Reciprocal altruism in Yanomamö food exchange. In *Adaptation and Human Behavior: An Anthropological Perspective*, edited by L. Cronk, N. Chagnon, and W. Irons, pp. 397-416. Hawthorne, NY: Aldine de Gruyter.

Hames, Raymond, and Carl McCabe (2007). Meal sharing among the Ye'kwana: An evaluation and refinement of some models. *Human Nature* 18: 1-21.

Hames, Raymond, and Patricia Draper (2004). Women's work, child care and helpers at the nest in a hunter-gatherer society. *Human Nature* 15(4): 319-341.

Harris, Grant T., N. Zoe Hilton, Marnie E. Rice, Angela W. Eke (2007). Children killed by genetic parents versus stepparents. *Evolution and Human Behavior* 28(2): 85-95.

Hawkes, K (2004). Mating, parenting and the evolution of human pair bonds. In *Kinship and Behavior in Primates*, edited by Bermard Chapais and Carol M. Berman, pp. 443-474. Oxford University Press.

Hawkes, K. (2000). Big game hunting and the evolution of egalitarian societies: Lessons from the Hadza. In *Hierarchies in Action: Cui Bono?*, edited by M. Diehl. Center for Archaeological Investigations, Occasional Paper 27: 59-83. Carbondale: Southern Illinois University Press.

Hawkes, K. (2002). Is meat the hunter's property? Ownership and explanations of hunting and sharing. In *Meat-eating and Human Evolution*, edited by C. Stanford and H. Bunn, pp. 219-236. Oxford: Oxford University Press.

Hawkes, K., and R. Bliege Bird (2002). Showing off, handicap signaling, and the evolution of men's work. *Evolutionary Anthropology* 11: 58-67.

Hawkes, K., J. F. O'Connell, and N.G. Blurton Jones (2002). Hunting and nuclear families: Some lessons from the Hadza about men's work. *Current Anthropology* 42(5): 681-709.

Hawkes, K., J. F. O'Connell, N. G. Blurton Jones, H. Alvarez and E. L. Charnov (2000). The grandmother hypothesis and human evolution. In *Adaptation and Human Behavior: An Anthropological Perspective*, edited by L. Cronk, N. Chagnon and W. Irons, pp. 231-252. New York: Aldine de Gruyter.

Hawkes, Kristen (2003). Grandmothers and the evolution of human longevity. *American Journal of Human Biology* 15: 380-400.

Hawkes, Kristen and James F. O'Connell (2005). How old is human longevity? *Journal of Human Evolution* 49: 650-653.

Hawkes, Kristen, and Nicholas Blurton Jones (2005). Human age structures, paleodemography, and the grandmother hypothesis. In *Grandmotherhood: The Evolutionary Significance of the Second Half of the Female Life*. Edited by E. Voland, C. Athanasios, and W. Schiefenhovel. pp. 118-142. Piscataway: Rutgers University Press.

Hayward, Lisa S., and Sievert Rohwer (2004). Sex difference in attitudes toward paternity testing. *Evolution and Human Behavior* 25(4): 242-248.

Henrich, Joseph, and Robert Boyd (2002). On modeling cognition and culture: Why replicators are not necessary for cultural evolution. *Journal of Cognition and Culture* 2(2): 87-112.

Henrich, Joseph, and Robert Boyd (2002). Why people punish defectors: Conformist transmission stabilizes costly enforcement of norms in cooperative dilemmas. *Journal of Theoretical Biology* 208: 79-89.

Hill, E.M., and Chow, K.L (2002). Life history theory and risky drinking. *Addiction* 97: 401-413.

Hill, K (2002). Cooperative food acquisition by Ache foragers. *Human Nature* 13(1): 105-128.

Hill, K., and J. Padwe (2000). Sustainability of Ache hunting in the Mbaracayu Reserve, Paraguay. In *Sustainability of Hunting in Tropical Forests*, J. Robinson and E. Bennet, eds., pp. 79-105. New York: Columbia University Press.

Hill, K., and M. Gurven (2004). Economic experiments to examine fairness and cooperation among the Ache Indians of Paraguay. In *Foundations of Human Sociality: Ethnography and Experiments in 15 Small-scale Societies*. J. Henrich, R. Boyd, S. Bowles, H. Gintis, E. Fehr, and Colin Camerer (eds.). Oxford University Press.

Hoagland, Mahlon (2001). *The Way Life Works*, reprint edition, Jones and Bartlett Publishers inc.

Horn, Toby M. (1993) Working with DNA and Bacteria in Precollege Science Classrooms. Reston, VA: National Association of Biology Teachers.

Hornyhead Branch restoration project . (n.d.). Retrieved September 11, 2006, from University of Tennessee Agricultural Experiment Station, Forest Resources Research and Education Center Web site at http://forestry.tennessee.edu/CUhornyhead.htm

Howard / White Unit No. 1 oil spill NRDA Obed Wild and Scenic River, Morgan County, Tennessee . Prepared for: National Park Service, Obed Wild and Scenic River, and Environmental Response, Planning, and Assessment Unit. Columbia: Research Planning, Inc. April 2003.

Hrdy, Sarah B (2000). The optimal number of fathers: Evolution, demography, and history in the shaping of female mate preferences. *Annals of New York Academy of Sciences* 907: 75-96.

Hrdy, Sarah B (2002). Motherhood. In *Encyclopedia of Evolution*. Mark Pagel, ed., pp. 55-65. Oxford: Oxford University Press.

Hrdy, Sarah B (2003). Maternal sentiments: How strong are they? Comment in reply to Arthur Wolf. *Current Anthropology* 44 (Supplement): S41-42.

Hrdy, Sarah B (2005). Comes the child before man: How cooperative breeding and prolonged postweaning dependence shaped human potential. In *Hunter-Gatherer Childhoods: Evolutionary, Developmental, and Cultural Perspectives*, B.S. Hewlett and M.E. Lamb, eds. pp. 65-91. New Brunswick: AldineTransaction.

Hrdy, Sarah B (2005). Cooperative breeders with an ace in the hole. In *Grandmotherhood: The Evolutionary Significance of the Second Half of the Female Life*. Edited by E. Voland, C. Athanasios, and W. Schiefenhovel. Piscataway: Rutgers University Press.

Hruschka, Daniel J., Daniel H. Lende, and Carol M. Worthman (2005). Biocultural dialogues: Biology and culture in psychological anthropology. *Ethos* 33(1): 1-19.

Hrushka, Daniel, and Joseph Henrich (2006). Friendship, cliquishness, and the emergence of cooperation. *Journal of Theoretical Biology* 239(1): 1-15.

http://www.tn-afs.org/Newsletter/2006%20TNAFS%20Newsletter.pdf

Huber, Brad R., Vendula Linhartova, and Dana Cope (2004). Measuring paternal certainty using cross-cultural data. *World Cultures* 15(1): 48-59.

Huber, Brad R., Vendula Linhartova, Dana Cope, and Mike Lacy (2004). Evolutionary theory and birth-related investments by kin in cross-cultural perspective. *World Cultures* 15(1): 60-78.

Hurtado A. M., K. Hill, H. Kaplan, and J. Lancaster (2002). The epidemiology of infectious diseases among South American Indians. *Current Anthropology* 42(3): 425-432.

Irons, William (2005). How has evolution shaped human behavior? Richard Alexander's contribution to an important question. *Evolution and Human Behavior* 26(1): 1-9.

Ivey, Paula K (2000). Cooperative reproduction in Ituri Forest hunter-gatherers: Who cares for Efe infants? *Current Anthropology* 41(5): 856-866.

Jamieson, Cheryl Sorenson, Laurel L Cornell, Paul L. Jamison, and Hideki Nakazato (2002). Are all grandmothers equal? A review and a preliminary test of the "grandmother hypothesis" in Tokugawa Japan. *American Journal of Physical Anthropology* 119: 67-76.

Jamison, Cheryl Sorenson, Paul L. Jamison and Laurel L. Cornell (2005). Human female longevity: How important is being a grandmother? In *Grandmotherhood: The Evolutionary Significance of the Second Half of the Female Life*. Edited by E. Voland, C. Athanasios, and W. Schiefenhovel. pp. 99-117. Piscataway: Rutgers University Press.

Jankowiak, William, and Monique Diderich (2000). Sibling solidarity in a polygamous community in the USA: Unpacking inclusive fitness. *Evolution and Human Behavior* 21(2): 125-139.

Janson, Charles H., and Eric A. Smith (2003). The evolution of culture: New perspectives and evidence. *Evolutionary Anthropology* 12(2): 57-60.

Jasienska, Grazyna, Ilona Nenko, and Michal Jasienski (2006). Daughters increase longevity of fathers, but daughters and sons equally reduce longevity of mothers. *American Journal of Human Biology* 18: 422-425.

Jensen, J. (1992) "Karyotyping with Hypercard." The American Biology Teacher, Vol. 54, (April).

JKaplan, Hillard, Jane Lancaster, W. Troy Tucker, and Kermyt G. Anderson (2002). Evolutionary approach to below replacement fertility. *American Journal of Human Biology* 14: 233-256.

Johnson, C. David, Timothy A. Kohler, and Jason Cowan (2005). Modeling Historical Ecology, Thinking about Contemporary Systems. *American Anthropologist* 107:96-108.

Johnson, Dominic D. P (2005). God's punishment and public goods: A test of the Supernatural Punishment Hypothesis in 186 world cultures. *Human Nature* 16(4): 410-446.

Johnson, George B. (2005). *Biology, Visualizing Life*. Holt, Rinehart, and Winston.

Kaplan, Hillard (2002). Human life history evolution. In *The Encyclopedia of Evolution*, M Pagel (cd). Oxford: Oxford University Press.

Kaplan, Hillard (2003). Evolutionary demography. In *The Encyclopedia of Population*, P. Demeny and G. McNicoll (eds). MacMillan: New York.

Keller, Matthew C., Randolph M. Nesse and Sandra Hofferth (2002). The Trivers-Willard hypothesis of parental investment: No effect in the contemporary United States. *Evolution and Human Behavior* 22(5): 343-360.

Kelly, Robert L., Lin Poyer, and Bram Tucker (2005). An ethnoarchaeological study of mobility, architectural investment, and food sharing among Madagascar's Mikea. *American Anthropologist* 107: 403-416.

Kennedy, G.E (2005). From the ape's dilemma to the weanling's dilemma: Early weaning and its evolutionary context. *Journal of Human Evolution* 48(2): 123-145.

Kennett, D.J., Atholl Anderson and B. Winterhalder (2006). The ideal free distribution, food production, and the colonization of Oceania. In *Behavioral Ecology and the Transition to Agriculture*, D.J. Kennett and B. Winterhalder, eds., pp. 265-288. Berkeley: University of California Press.

Kennett, Douglas J., and Bruce Winterhalder (2006). *Behavioral Ecology and the Transition to Agriculture*. Berkeley: University of California Press.

Kohler, Timothy A., and Sander van der Leeuw (editors) (2007). *The Model-Based Archaeology of Socionatural Systems*. SAR Press, Santa Fe.

Kramer, Karen L (2002). Variation in juvenile dependence: Helping behavior among Maya children. *Human Nature* 13: 299-325.

Kramer, Karen L (2004). Reconsidering the cost of childbearing: The timing of children's helping behavior across the life cycle of Maya families *Research in Economic Anthropology* 23: 335-353.

Kramer, Karen L (2005). Children's help and the pace of reproduction: Cooperative breeding in humans. *Evolutionary Anthropology* 14(6): 224-237.

Kramer, Karen L (2005). *Maya Children. Helpers at the Farm*. Cambridge: Harvard University Press.

Kramer, Karen L., and Garnett P. McMillan (2006). The effect of labor-saving technology on longitudinal fertility changes. *Current Anthropology* 47(1): 165-172.

Kramer, Karen L., and James L. Boone (2002). Why intensive agriculturalists have higher fertility: A household labor budget approach to subsistence intensification and fertility rates. *Current Anthropology* 43(3): 511-517.

Kramer, Karen. Generational transfers across the life cycle of the family: A cross-cultural perspective.

Kramer, Patricia Ann (2004). Burden transport: When, how and how much? *Research in Economic Anthropology* 23: 249-269.

Kruger, Daniel J., and Randolph M. Nesse (2006). An evolutionary life-history framework for understanding sex differences in human mortality rates. *Human Nature* 17: 74-97.

Kuznar, Lawrence A (2002). Evolutionary applications of risk sensitivity models to socially stratified species: Comparison of sigmoid, concave, and linear functions. *Evolution and Human Behavior* 23(4): 265-280.

Lahdenperä, M.,V. Lummaa, and A.F. Russell (2004). Menopause: why does fertility end before life? *Climacteric* 7: 327-331

Lam, David (2003). Evolutionary biology and rational choice in models of fertility. In *The Biodemography of Human Reproduction and Fertility*, J. Rodgers and H.P. Kohler, eds., pp. 322-338. Boston: Kluwer Academic Publishers.

Lancaster, J., H. Kaplan, K. Hill, and A.M. Hurtado (2000). The evolution of life history, intelligence and diet among chimpanzees and human foragers. In *Evolution, Culture and Behavior, Perspectives in Ethology*, Volume 13, eds. F. Tonneau and N. S. Thompson. New York: Plenum.

Lancaster, Jane B., and Hillard S. Kaplan (2000). Parenting other men's children: Cost, benefits, and consequences. In *Adaptation and Human Behavior: An Anthropological Perspective*, edited by L. Cronk, N. Chagnon, and W. Irons, pp. 179-201. Hawthorne, NY: Aldine de Gruyter.

Leonetti, Donna L., Dilip C. Nath, Natabar S. Hemam, and Dawn B. Neill (2005). Kinship organization and grandmother's impact on reproductive success among the matrilineal Khasi and patrilineal Bengali of N.E. India. In *Grandmotherhood: The Evolutionary Significance of the Second Half of Female Life*. E. Voland, A. Chasiotis, and W. Schiefenhoevel (eds.) pp. 194-294. Rutgers University Press.

Leslie, P. and B. Winterhalder (2002). Demographic consequences of unpredictability in fertility outcomes. *American Journal of Human Biology* 14: 168-183.

Locatelli, Francesco; Aljama, Pedro; Barany, Peter; et al. (2004). "Erythropoiesis-Stimulating Agents and Antibody-Mediated Pure Red-Cell Aplasia: Where Are We Now and Where Do We Go From Here?," *Nephrol Dial Transplant*. 19: 288-293.

Low, Bobbi (2005). Families: An evolutionary ecological approach. In *Families in Global Perspective*, edited by J. L. Roonarine and U. Gielen. Boston: Allyn & Bacon.

Low, Bobbi S (2005). George Williams: Theoretician and guerilla environmentalist. *Quarterly Review of Biology* 80(1): 72-83.

Low, Bobbi S., Carl P. Simon, and Kermyt G. Anderson (2002). An evolutionary ecological perspective on demographic transitions: Modeling multiple currencies. *American Journal of Human Biology* 14: 149-167.

Low, Bobbi S., Carl P. Simon, and Kermyt G. Anderson (2003). The biodemography of modern women: Tradeoffs when resources become limiting. In *The Biodemography of Human Reproduction and Fertility*, J. Rodgers and H.P. Kohler, eds., pp. 105-134. Boston: Kluwer Academic Publishers.

Lummaa, V (2003). Reproductive success and early developmental conditions in humans: Downstream effects of prenatal famine, birth weight and timing of birth. *American Journal of Human Biology* 15: 170-179.

Lummaa, V., and M. Tremblay (2003). Month of birth predicted reproductive success and fitness in pre-modern Canadian women. *Proceedings of the Royal Society of London B* 270: 2355-2361.

Lummaa, V., and T.H. Clutton-Brock (2002). Early development, survival and reproduction in humans. *Trends in Ecology and Evolution* 17: 141-147.

Lummaa, V., J. Jokela, and E. Haukioja (2002). Gender difference in benefits of twinning in pre-industrial humans: Boys did not pay. *Journal of Animal Ecology* 70: 739-746.

Lynn, Richard; Van Court, Marilyn (2004). "New evidence of dysgenic fertility for intelligence in the United States". *Intelligence* 32 (2): p. 193. Ablex Pub.

Mace, Ruth, and Jennifer Eardley (2004). Maternal nutrition and sex ratio at birth in Ethiopia. *Research in Economic Anthropology* 23: 295-306.

Maestripieri, Dario, and Suzanne Pelka (2002). Sex differences in interest in infants across the lifespan: A biological adaptation for parenting? *Human Nature* 13: 327-344.

Magner, LN (2002). *A History of the Life Sciences*. TF-CRC.

Marlowe, F.W., C.L. Apicella, and D. Reed (2005). Men's preferences for women's profile waist-hip-ratio in two societies. *Evolution and Human Behavior* 26: 458-468.

Marlowe, Frank (2002). Male contribution to diet and female reproductive success among foragers. *Current Anthropology* 42(5): 755-760.

Marlowe, Frank (2003). A critical period for provisioning by Hadza men: Implications for pair bonding. *Evolution and Human Behavior* 24(3): 217-229.

Marlowe, Frank (2005). Who tends Hadza children? In *Hunter-Gatherer Childhoods: Evolutionary, Developmental, and Cultural Perspectives*, B.S. Hewlett and M.E. Lamb, eds. pp. 177-190. New Brunswick: AldineTransaction.

Micklos, D.A. and Freyer, G. A. (1990) DNA Science: A First Course in Recombinant DNA Technology. Cold Spring Harbor Laboratory Press and Carolina Biological Supply Company.

Morange, M & Cobb, M (2000). *A History of Molecular Biology*. Harvard University Press.

Morange, M & Cobb, M (2000). A History of Molecular Biology. Harvard University Press.

MS 0828: The Tennessee Valley Authority reports, 1933-1973 . (n.d.). Retrieved September 11, 2006, from http://www.lib.utk.edu/spcoll/manuscripts/ms0828fa.html

Mueller, Ulrich (2004). Does late reproduction extend the life span? Findings from European royalty. *Population and Development Review* 30(3): 449–466.

Mulcahy, Henry (1989) "A Simulated Laboratory Exercise in Genetic Engineering." Abstracts of the Annual Meeting of American Society for Microbiologists.

Mullis, Kary B. (1990) "The Unusual Origin of the Polymerase Chain Reaction." Scientific American. April: 56-61.

Nagaoka, Lisa (2002). Explaining subsistence change in southern New Zealand using foraging theory models. *World Archaeology* 34(1): 84-102.

Nagaoka, Lisa (2002). The effects of resource depression on foraging efficiency, diet breadth, and patch use in southern New Zealand. *Journal of Anthropological Archaeology* 21(4): 419-442.

Nath, D. C., D. L. Leonetti, and M.S. Steele (2000). Analysis of birth intervals in a non-contracepting Indian population: An evolutionary ecological approach. *J. Biosocial Sci.* 32: 343-354.

National Park Service. (2005, January 4). *Obed National Wild and Scenic River* . Retrieved September 11, 2006, from http://www2.nature.nps.gov/geology/parks/obri/index.cfm

National Parks Directory. (n.d.). *Obed Wild and Scenic River — Tennessee* . Retrieved December 7, 2006, from http://classic.mountainzone.com/nationalparks/obed/index.html

Nettle, D (2002). Height and reproductive success in a cohort of British men. *Human Nature* 13(4): 473-491.

Noss, Andy and Barry S. Hewlett (2002). The contexts of female hunting in central Africa. *American Anthropologist* 103: 1024-1040.

O'Connell, J. F (2000). An emu hunt. In *Australian Archaeologist: Collected Papers in Honour of Jim Allen*, edited by A. Anderson and T. Murray, pp. 172-181. Division of Archaeology and Natural History, Research School of Pacific and Asian Studies. Canberra: Australian National University.

O'Connell, J. F., K. Hawkes, and N. G. Blurton Jones (2002). Meat-eating, grandmothering and the evolution of early human diets. In *Human Diet: Its Origin and Evolution*, edited by P. Unger and M. Teaford, pp. 49-60. Westport, Conn.: Bergin and Garvey.

O'Connell, J. F., K. Hawkes, K. Lupo, and N. G. Blurton Jones (2002). Male strategies and Plio-Pleistocene archaeology. *Journal of Human Evolution* 43: 831-872.

O'Connor, C. M (2004). Effects Of central decisions On local livelihoods in Indonesia: Potential synergies between the programs of transmigration and industrial forest conversion. *Population and Environment* 25(4): 319-333.

Old, R.W. and Primrose, S.B. (1989) Principles of Gene Manipulation 4th Edition. London: Blackwell Scientific Publications.

Ostermeier, D., Fly, M., Muth, A., Pavey, J., and Steiner, M. (n.d.). *Reluctant landowners: opening the door to collaborative forestry* . Retrieved December 18, 2006, from http://www.infor.cl/webinfor/eventos/IUFRO_Valdivia2002/textos/David%20Ostermeier%20et%20al.%20USA.pdf

Paciotti, B., C. Hadley, C. Holmes, and M. Borgerhoff Mulder (2005). Grass-roots justice in Tanzania. *American Scientist* 93: 58-64.

Paciotti. B., and C. Hadley (2004). Large-scale cooperation among Sungusungu "vigilantes" of Tanzania: Conceptualizing micro-economic and institutional approaches. *Research in Economic Anthropology* 23: 119-147.

Panicali, D. and Paoletti, E. (1982) "Construction of pox viruses as cloning vectors: Insertion of the thymidine kinase gene from herpes simplex virus into the DNA of infectious vaccinia virus." Proc. Natl. Acad. Sci. 79:4927-4931.

Parsons, J.W., and Benson, N.G. (1960). Fertilization of Obed River, Tennessee . *Journal of the Tennessee Academy of Science*, 3(1), 63-75.

Peccei, Jocelyn Scott (2002). A critique of the grandmother hypotheses: Old and new. *American Journal of Human Biology* 13: 434-452.

Peccei, Jocelyn Scott (2005). Menopause: adaptation and epiphenomenon. In *Grandmotherhood: The Evolutionary Significance of the Second Half of the Female Life.* Edited by E. Voland, C. Athanasios, and W. Schiefenhovel. pp. 38-58. Piscataway: Rutgers University Press.

Pennington and Associates. (n.d.). *A survey of the Periphyton Community, Clear Creek, Tennessee* . For the National Park Service, Big South Fork National River and Recreation Area (BSFNRRA), Obed Wild and Scenic River.

Pennington, R (2002). Economic stratification and health among the Herero of Botswana. In W. R. Leonard and M. H. Crawford, eds., *Human Biology of Pastoral Populations*, pp. 183-205. Cambridge University Press.

Pennington, R (2002). Hunter-gatherer demography. In C. Panter-Brick, R. Layton and P. Rowley-Conwy, eds., *Hunter-gatherers: An Interdisciplinary Perspective*, pp. 170-204. Cambridge University Press.

Pettay, J.E., L.E.B. Kruuk, J. Jokela, and V. Lummaa (2005). Heritability and genetic constraints of life-history trait evolution in preindustrial humans. *Proceedings of the National Academy of Science of U.S.A.* 102: 2838-2843.

Pike, I. L (2000). The nutritional consequences of pregnancy sickness: A critique of a hypothesis. *Human Nature* 11: 207-232.

Pike, Ivy L (2005). Maternal stress and fetal responses: Evolutionary perspectives on preterm delivery. *American Journal of Human Biology* 17(1): 55-65.

Pike, Ivy L., and Sharon R. Williams (2006). Incorporating psychosocial health into biocultural models: Preliminary findings from Turkana women of Kenya. *American Journal of Human Biology* 18(6): 729-740.

Piperno, Dolores R (2006). The origins of plant cultivation and domestication in the neotropics: A behavioral ecological perspective. In *Behavioral Ecology and the Transition to Agriculture*, D.J. Kennett and B. Winterhalder, eds., pp. 137-166. Berkeley: University of California Press.

Polioudakis, Emanuel (2000). Relatedness, class, and social organization in a village in southern Thailand. *Evolution and Human Behavior* 21(5): 297-316.

Pollet, Thomas V (2007). Genetic relatedness and sibling relationship characteristics in a modern society. *Evolution and Human Behavior* 28(3): 176-185.

Porter, C.C., and Marlowe, F.W (2007). How marginal are forager habitats? *Journal of Archaeological Science* 34:59-68.

Price ME (2006). Judgments about cooperators and freeriders on a Shuar work team: An evolutionary psychological perspective. *Organizational Behavior and Human Decision Processes* 101: 20-35.

Price, M.E (2003). Pro-community altruism and social status in a Shuar village. *Human Nature* 14: 191-208.

Price, M.E., Cosmides, L., and Tooby, J (2002). Punitive sentiment as an anti-free rider psychological device. *Evolution and Human Behavior* 23: 203-231.

Price, ME (2006). Monitoring, reputation and "greenbeard" reciprocity in a Shuar work team. *Journal of Organizational Behavior* 27: 201-219.

Price, Michael E (2005). Punitive sentiment among the Shuar and in industrialized societies: cross-cultural similarities. *Evolution and Human Behavior* 26(3): 279-287.

Quinlan, R (2002). Effect of household structure on female reproductive strategies in a Caribbean community. *Human Nature* 12(3): 169-189.

Quinlan, R (2003). Father-absence, parental care and female reproductive development. *Evolution and Human Behavior* 24(6): 376-390.

Quinlan, R (2006). Gender and risk in a matrifocal Caribbean community: A view from behavioral ecology. *American Anthropologist* 108(3): 469-79.

Quinlan, R (2007). Human parental effort and environmental risk. *Proceedings of the Royal Society B: Biological Sciences*. 274(1606):121-125.

Quinlan, R. 2005 Kinship, gender & migration from rural Dominica. *Migration Letters* 2(1): 2-12.

Quinlan, R., and M. Flinn (2003). Intergenerational transmission of conjugal stability in a Caribbean village. *Journal of Comparative Family Studies* 34: 569-583.

Quinlan, R., Quinlan, M., and Flinn, M (2003). Parental investment and age at weaning in a Caribbean village. *Evolution and Human Behavior* 24: 1-16.

Quinlan, R.J., M.B. Quinlan, and M.V. Flinn (2005). Local resource enhancement and sex-biased breastfeeding in a Caribbean community. *Current Anthropology* 46(3): 471-480.

Quinlan, Robert, and Mark V. Flinn (2005). Kinship, sex and fitness in a Caribbean community. *Human Nature* 16(1): 32-57.

Raleigh, R.F., and Nelson, P.C. (1985). *Habitat suitability, index models and instream flow suitability curves: pink salmon* (pdf file). Report no. 82(10.109). Fort Collins: Western Energy and Land Use Team, U.S. Fish and Wildlife Service.

Ranta, E., V. Lummaa, V. Kaitala, and J. Merilä (2000). Spatial dymamics of adaptive sex ratios. *Ecology Letters* 3: 30-34.

Rasmussin and Matheson (eds) (1990) A Sourcebook of Biotechnology Activities. National Association of Biology Teachers, 11250 Roger Bacon Drive #19, Reston, VA 22090.

Reason, Letitia L (2004). The behavioral ecology of female genital cutting in northern Ghana. *Research in Economic Anthropology* 23: 175-202.

Rende Taylor, Lisa (2004). Maintaining the matriline: Children's birth order roles and school drop-out among Thai Khon Muang. *Research in Economic Anthropology* 23: 355-377.

Rende Taylor, Lisa (2005). Dangerous trade-offs: The behavioral ecology of child labor and prostitution in rural northern Thailand. *Current Anthropology* 46(3): 411-431.

Rende Taylor, Lisa (2005). Patterns of child fosterage in rural northern Thailand. *Journal of Biosocial Science* 37(3): 333-350.

Richerson, Peter J., and Robert Boyd (2000). Climate, culture, and the evolution of cognition. In C.M. Heyes and L. Huber (eds.) *The Evolution of Cognition*, pp. 329-345. Cambridge: MIT Press.

Richerson, Peter J., and Robert Boyd (2000). Evolution: The Darwinian theory of social change—An homage to Donald T. Campbell. In *Paradigms of Social Change* W. Schelkle, W-H. Krauth, M. Kohli, and G. Elwert (eds.), pp. 257-282. New York: Saint Martin's Press.

Richerson, Peter J., and Robert Boyd (2000). Memes: Universal acid or better mouse trap? In *Darwinizing Culture* (Robert Aunger, ed.), Cambridge, Cambridge University Press.

Richerson, Peter J., and Robert Boyd (2000). The Pleistocene and the origins of human culture: Built for speed. *Perspectives in Ethology* 13: 1-45.

Richerson, Peter J., and Robert Boyd (2002). Built for speed, not for comfort: Darwinian theory and human culture. *History and Philosophy of the Life Sciences* 23: 423-463.

Richerson, Peter J., and Robert Boyd (2002). Institutional evolution in the Holocene: The rise of complex societies. In *The Origin of Human Social Institutions*, pp. 197-204, W.G. Runciman, ed. Proceedings of the British Academy 110.

Richerson, Peter J., and Robert Boyd (2002). The evolution of subjective commitment to groups: A tribal instincts hypothesis. In *The Evolution and the Capacity for Subjective Commitment*, R.M. Nesse (ed.), pp. 186-220, Russell Sage.

Richerson, Peter J., Robert Boyd, and Brian Paciotti (2002). An evolutionary theory of commons management. In *The Drama of the Commons*, Elinor Ostrom, Thomas Dietz, Nives Dolak, Paul C. Stern, Susan Stonich, and Elke U. Weber, eds., pp. 403-442. National Academy Press.

Richerson, Peter J., Robert Boyd, and Robert L. Bettinger (2002). Was agriculture impossible during the Pleistocene but mandatory during the Holocene? A climate change hypothesis. *American Antiquity* 66: 387-411.

Richerson, Peter, Robert Boyd, and Joseph Henrich (2003). The cultural evolution of cooperation. In *Genetic and Culture Evolution of Cooperation*, edited by Peter Hammerstein. Cambridge, MA: MIT Press.

Riddle, J.W., Jr. (1975, June). *Status of the native muskellunge,* Esox masquinongy ohioensis, *of the Cumberland Plateau, Tennessee* . Master's Thesis. Tennessee Technological University: Cookeville.

Robson, Shannen L (2004). Breast milk, diet, and large human brains. *Current Anthropology* 45: 419-425.

Robson, Arthur J., and Hillard S. Kaplan (2003). The evolution of human life expectancy and intelligence in hunter-gatherer economies. *American Economic Review* 93(1): 150-169.

Rodseth, L. T., and Shannon A. Novak (2000). The social modes of men: Toward an ecological model of human male relationships. *Human Nature* 11(4): 335-366.

Roth, Eric Abella, Elizabeth Ngugi, and Masako Fujita (2006). Self-deception does not explain high-risk sexual behavior in the face of HIV/AIDS: A test from northern Kenya. *Evolution and Human Behavior* 27(1): 53-62.

Russ, W.T. (2006). *Current distribution and seasonal habitat use of the threatened spotfin chub in the Emory River Watershed* (pdf file). Master's Thesis. Tennessee Technological University: Cookeville.

Ruttan, Lore M (2003). Finding fish: Grouping and catch per unit effort in the Pacific Hake (*Merluccius productus*) fishery. *Canadian Journal of Fisheries and Aquatic Sciences* 60: 1068-1077.

Ruttan, Lore M (2006). Sociocultural heterogeneity and the commons. *Current Anthropology* 47(5): 843-853.

Saha, Atanu; Grabowski, Henry; Birnbaum, Howard; Greenberg, Paul; Bizan, Oded (2006). "Generic Competition in the U.S. Pharmaceutical Industry," *International Journal of the Economics of Business*, 13(1): 15-38(24)

Sasser, M.A. (2003, May). *Indicator bacteria levels in the waters of two east Tennessee watersheds* . Master's Thesis. University of Tennessee: Knoxville.

Schank, Jeffrey C (2006). Do human menstrual-cycle pheromones exist? *Human Nature* 17(4): 449-470.

Schellekens H. (2004). "Biosimilar Therapeutic Agents: Issues with Bioequivalence and Immunogenicity," *Eur J Clin Invest.* 34(12):797-799.

Schellekens, Huub (2002). "Bioequivalence and the Immunogenicity of Biopharmaceuticals," *Nature Reviews*; 1: 457-462

Schiefenhovel, Wulf, and Andreas Grabolle (2005). The role of maternal grandmothers in Trobriand adoptions. In *Grandmotherhood: The Evolutionary Significance of the Second Half of the Female Life*. Edited by E. Voland, C. Athanasios, and W. Schiefenhovel. pp. 177-214. Piscataway: Rutgers University Press.

Schmalzer, P.A. (1982). *Vegetation of the Obed Wild and Scenic River, Tennessee, and a comparison of reciprocal averaging ordination and binary discriminant analysis* Doctoral Dissertation. University of Tennessee: Knoxville.

Schölmerich, Axel, Birgit Leyendecker, Banu Citlak-Kisagün, Amy Miller, and Robin Harwood (2005). Variability of the role of grandmothers. In *Grandmotherhood: The Evolutionary Significance of the Second Half of the Female Life*. Edited by E. Voland, C. Athanasios, and W. Schiefenhovel. Piscataway: Rutgers University Press.

Sear R., Mace R., and I. A. McGregor (2000). Maternal grandmothers improve nutritional status and survival of children in rural Gambia. *Proc. Roy. Soc. Lond. B.*

Sear, R., Mace, R., Shanley, D., and McGregor, I.A (2002). The fitness of twin mothers: Evidence from rural Gambia. *Journal of Evolutionary Biology* 14: 433-443.

Sear, R., N. Allal, and R. Mace (2004). Height, marriage and reproductive success in Gambian women. *Research in Economic Anthropology* 23: 203-224.

Sear, Rebecca (2006). Height and reproductive success: How a Gambian population compares with the West. *Human Nature* 17(4): 405-418.

Sear, Rebecca (2007). The impact of reproduction on Gambian women: Does controlling for phenotypic quality reveal costs of reproduction? *American Journal of Physical Anthropology* 132 (4): 632-641.

Sear, Rebecca, Fiona Steele, Ian A. McGregor, and Ruth Mace (2002). The effects of kin on child mortality in rural Gambia. *Demography* 39(1): 43-63.

Sear, Rebecca, Ruth Mace, and Ian A. McGregor (2003). A life-history analysis of fertility rates in rural Gambia: Evidence for trade-offs or phenotypic correlations? In *The Biodemography of Human Reproduction and Fertility*, J. Rodgers and H.P. Kohler, eds., pp. 135-160. Boston: Kluwer Academic Publishers.

Sear, Rebecca, Ruth Mace, and Ian A. McGregor (2003). The effects of kin on female fertility in rural Gambia. *Evolution and Human Behavior* 24(1): 25-42.

Sell, A., E.H. Hagen, L. Cosmides, and J. Tooby (2003). Evolutionary theory: Implications for Cognitive Science. In *Encyclopedia of Cognitive Science Nature Publishing Group*, v. 2, pp. 47-53. Macmillan Publishers Ltd.

Sellen, D. W (2000). Relationships between fertility, mortality and subsistence: Results of recent phylogenetic analyses. In *Humanity from African Naissance to Coming Millennia*. P.V. Tobias, M. Raath, J. Moggi-Cecchi, G. Doyle eds. Florence University Press, Firenze and Witwatersrand University Press, Johannesburg, pp. 51-64.

Sellen, D. W (2002). Anthropological approaches to understanding variation in breastfeeding and application to promotion of "Baby Friendly" communities. *Nutritional Anthropology* 25(1): 19-29.

Sellen, D. W (2002). Comparison of infant feeding patterns reported for non-industrial populations with current recommendations. *Journal of Nutrition* 131(10): 2707-2715.

Sellen, D. W (2002). Of what use is an evolutionary anthropology of weaning? *Human Nature* 12(1): 1-7.

Sellen, D. W (2002). Sub-optimal breast feeding practices: Ethnographic approaches to building "Baby Friendly" communities. In *Integrating Population Outcomes, Biological Mechanisms, and Research Methods in the Study of Human Milk and Lactation* (Advances in Experimental Medicine and Biology Vol. 503). M. Davis, L. Hanson, C. Isaacs and A. Wright, eds. New York, Kluwer Academic/Plenum Publishers, pp. 223-232.

Sellen, D. W (2002). Weaning, complementary feeding and maternal decision making in a rural east African pastoral population (Datoga). *Journal of Human Lactation* 17(3): 233-244.

Sellen, D. W (2003). Nutritional consequences of wealth differentials in East African pastoralists: The case of the Datoga of Northern Tanzania. *Human Ecology* 31(4): 529-570.

Sellen, D. W., and D. B. Smay (2002). Relationship between subsistence and age at weaning in "pre-industrial" societies. *Human Nature* 12(1): 47-87.

Sellen, D. W., M. Borgerhoff Mulder, and D.F.S. Sieff (2000). Fertility, offspring quality and wealth in Datoga pastoralists: Testing evolutionary models of intersexual selection. In *Adaptation and Human Behavior: An Anthropological Perspective*, edited by L. Cronk, N. Chagnon, and W. Irons, pp. 91-114. Hawthorne, NY: Aldine de Gruyter.

Sellen, Daniel W., and Daniel J. Hruschka (2004). Extracted-food resource-defense polygyny in native western North American societies at contact. *Current Anthropology* 45(5): 707-714.

Shenk, Mary (2005). Kin investment in wage-labor economies: Effects on child and marriage market outcomes. *Human Nature* 16: 81-113.

Shenk, Mary K (2004). Embodied capital and heritable wealth in complex cultures: A class-based analysis of parental investment in urban south India. *Research in Economic Anthropology* 23: 307-333.

Sherman, Paul W., and Geoffrey A. Hash (2002). Why vegetable recipes are not very spicy. *Evolution and Human Behavior* 22(3): 147-163.

Sherry, D.S., and Marlowe, F.W (2007). Anthropometry shows nutritional homogeneity among Hadza foragers. *American Journal of Human Biology* 19:107-118.

Shropshire, William O (2003). An economic approach to the evolution of male-female exchange. *Human Nature* 14: 235-266.

Sievert, Lynnette Leidy, and Catherine A. Dubois (2005). Validating signals of ovulation: Do women who think they know, really know? *American Journal of Human Biology* 17(3): 310-320.

Smith, Bruce D (2006). Human behavioral ecology and the transition to food production. In *Behavioral Ecology and the Transition to Agriculture*, D.J. Kennett and B. Winterhalder, eds., pp. 289-304. Berkeley: University of California Press.

Smith, E.A (2000). Three styles in the evolutionary analysis of human behavior. In *Adaptation and Human Behavior: An Anthropological Perspective*, edited by L. Cronk, N. Chagnon, and W. Irons, pp. 27-46. Hawthorne, NY: Aldine de Gruyter.

Smith, E.A, M. Borgerhoff-Mulder, and K. Hill (2002). Controversies in the evolutionary social sciences: A guide for the perplexed. *Trends in Ecology and Evolution* 16: 128-135.

Smith, E.A., and B. Winterhalder (2003). Human behavioral ecology. In *Encyclopedia of Cognitive Science*, Volume 2. Lynn Nadel, ed., pp. 377-385. London: Nature Publishing Group.

Smith, E.A., and R. Bliege Bird (2000). Turtle hunting and tombstone opening: Public generosity as costly signaling. *Evolution and Human Behavior* 21: 245-261.

Smith, E.A., R. Bliege Bird, and D.W. Bird (2003). The benefits of costly signaling: Meriam turtle hunters. *Behavioral Ecology* 14: 116-126.

Smith, Eric A., and B. Winterhalder (2002). Evolutionary social science: The behavioral ecology approach. *Samfundsøkonomen* 4: 13-21.

Smith, Eric Alden (2003). Human cooperation: Perspectives from behavioral ecology. In *The Genetic and Cultural Evolution of Cooperation*, ed. P. Hammerstein, pp. 401-427. Cambridge, MA: MIT Press.

Smith, Eric Alden (2006). Anthropological schisms. *Anthropology Newsletter*, January 2006, pp. 8-11.

Smith, R.P. (1979). *Status report on the occurrence of the spotfin chub,* Hyboposis monacha (Cope)*, in the Emory River, Morgan County, Tennessee* (pdf file). Gatlinburg: Fishery Resources.

Snyder, Karen (2004). Risk perception and resource security for female agricultural workers. *Research in Economic Anthropology* 23: 271-292.

Soltis, J., and R. McElreath (2002). Can females gain additional paternal investment by mating with multiple males? A game theoretic approach. *American Naturalist* 158(5): 519-529.

Sosis, R (2000). Costly signaling and torch fishing on Ifaluk Atoll. *Evolution and Human Behavior* 21: 223-244.

Sosis, R (2000). The emergence and stability of cooperative fishing on Ifaluk Atoll. In *Human Behavior and Adaptation: an Anthropological Perspective*, edited by L. Cronk, N. Chagnon, and W. Irons, pp. 437-472. Hawthorne, NY: Aldine de Gruyter.

Sosis, R., and E. Bressler (2003). Cooperation and commune longevity: A test of the costly signaling theory of religion. *Cross-Cultural Research* 37: 211-239.

Sosis, Richard (2004). The adaptive value of religious ritual. *American Scientist* 92: 166-172.

Sosis, Richard, and Bradley J. Ruffle (2004). Ideology, religion, and the evolution of cooperation: field experiments on Israeli kibbutzim. *Research in Economic Anthropology* 23: 89-117.

Southern Division of the American Fisheries Society 2006 Spring Meeting program . (2006, February). Retrieved December 5, 2006, from http://www.sdafs.org/meetings/2006/2006program.pdf

Stedman, S., and Stedman, B. (2002). *Notes on the birds of the Big South Fork National River and Recreation Area and the Obed River National Wild and Scenic River* . Cookeville: Tennessee Technological University.

Stein, A.D., Barnett P.G., and D.W. Sellen (2004). Maternal undernutrition and the sex ratio at birth in Ethiopia: Evidence from a national sample. *Biology Letters* 271(S3): S37- S39.

Stephen C. Stearns (2005). Issues in evolutionary medicine. *American Journal of Human Biology* 17(2): 131 – 140.

Stevenson, Joan C., Phillip M. Everson, and Michael Grimes (2004). Reproductive measures, fitness, and migrating Mennonites: An evolutionary analysis. *Human Biology* 76(5): 667-687.

Stewart-Williams, Steve (2007). Altruism among kin vs. nonkin: effects of cost of help and reciprocal exchange. *Evolution and Human Behavior* 28(3): 193-198.

Strassmann, B. I (2000). Polygyny, family structure, and child mortality: A prospective study among the Dogon of Mali. In *Adaptation and Human Behavior: An Anthropological Perspective*, edited by L. Cronk, N. Chagnon, and W. Irons, pp. 49-67. Hawthorne, NY: Aldine de Gruyter.

Strassmann, Beverly I., and Brenda Gillespie (2002). Life-history theory, fertility and reproductive success in humans. *Proceedings Of The Royal Society Of London Series B* 269(1491): 553-562.

Strassmann, Beverly I., and Brenda Gillespie (2003). How to measure reproductive success? *American Journal of Human Biology* 15: 361-369.

Sugiyama, L.S., and R. Chacon (2005). Juvenile responses to household ecology among the Yora of Peruvian Amazonia. In: Hewlett B, Lamb M. (Eds). *Hunter-Gatherer Childhoods: Evolutionary, Developmental, and Cultural Perspectives*, pp. 237-251. New York: Aldine.

Sugiyama, L.S., J. Tooby, and L. Cosmides (2002). Cross-cultural evidence of cognitive adaptations for social exchange among the Shiwiar of Ecuadorian Amazonia. *Proceedings of the National Academy of Sciences* 99(17): 11537-11545.

Sugiyama, L.S., J. Tooby, and L. Cosmides (2003). Cross-cultural evidence of cognitive adaptations for social exchange among the Shiwiar of Ecuadorian Amazonia. Reprinted in Kenrick DT, Luce CL, editors. *The functional mind: Readings in evolutionary psychology.* pp. 97-109. USA: Allen & Bacon.

Sugiyama, Lawrence S (2004). Illness, injury, and disability among Shiwiar forager-horticulturalists: Implications of health-risk buffering for the evolution of human life history. *American Journal of Physical Anthropology* 123: 371-389.

Sugiyama, Lawrence Scott, and Michelle Scalise Sugiyama (2003). Social roles, prestige, and health risk: Social niche specialization as a risk-buffering strategy. *Human Nature* 14: 165-190.

Sugiyama, Lawrence, and Richard Chacon (2000). Effects of illness and injury on foraging among the Yora and Shiwiar: Pathology risk as adaptive problem. In *Adaptation and Human Behavior: An Anthropological Perspective*, edited by L. Cronk, N. Chagnon, and W. Irons, pp. 371-395. Hawthorne, NY: Aldine de Gruyter.

Sullivan, R.J., and E.H. Hagen (2002). Psychotropic substance-seeking: Evolutionary pathology or adaptation? *Addiction* 97: 389-400.

Summers, K (2005). The evolutionary ecology of despotism. *Evolution and Human Behavior* 26(1): 106-135.

Tennessee Valley Authority. (1970, April). *Tennessee valley streams: Their fish, bottom fauna, and aquatic habitat* (pdf file). *The Emory River, 1968*. Division of Forestry, Fisheries and Wildlife Development.

Thorpe, Robin and Meenu, Wadhwa (2007). "Unwanted Immunogenicity: Implications for Follow-on Biologicals," *Drug Information Journal*. 41(01)

Ting, Tih-Fen (2004). Resources, fertility, and parental investment in Mao's China. *Population and Environment* 25(4): 281-297.

Ting, Tih-Fen (2004). Shifts in reproductive patterns in China. *Population and Environment* 25(4): 299-317.

Tooley GA, Karakis M, Stokes M, Ozanne-Smith J (2006). Generalizing the Cinderella Effect to unintentional childhood fatalities. *Evolution and Human Behavior* 27(3): 224-230.

Torrey, J.G. (1987). "The Development of Plant Biotechnology." American Scientist. 73:354-363.

Towner, Mary C (2002). Linking dispersal and marriage in humans: Life history data from Oakham, Massachusetts, USA (1750-1850). *Evolution and Human Behavior* 23(5): 337-357.

Towner, Mary C (2002). Linking dispersal and resources in humans: Life history data from Oakham, Massachusetts (1750-1850). *Human Nature* 12: 321-349.

Tracer, David (2004). Market integration, reciprocity and fairness in rural Papua New Guinea: Results from a two-village ultimatum game experiment. In *Foundations of Human Sociality: Ethnography and Experiments in 15 Small-scale Societies*. J. Henrich, R. Boyd, S. Bowles, H. Gintis, E. Fehr, and Colin Camerer (eds.). Oxford University Press.

Tracer, David P (2002). Somatic versus reproductive energy allocation in Papua New Guinea: Life history theory and public policy. *American Journal of Human Biology* 14: 621-626.

Tracer, David P (2003). Selfishness and fairness in economic and evolutionary perspective: An experimental economic study in Papua New Guinea. *Current Anthropology* 44: 423-438.

Trapper Arne's unique crayfish page: catch crayfish with traps from Trapper Arne . (n.d.). Retrieved December 5, 2006, from http://www.trapperarne.com/crayfishbystate.htm#topofpage

Tucker, Bram (2006). A future discounting explanation for the persistence of a mixed foraging-horticulture strategy among the Mikea of Madagascar. In *Behavioral Ecology and the Transition to Agriculture*, edited by D. J. Kennett and B. Winterhalder. Berkeley: University of California Press. Pp. 22-40.

Tucker, Bram, and Alyson G. Young (2005). Growing up Mikea: Children's time allocation and tuber foraging in southwestern Madagascar. In *Hunter-Gatherer Childhoods*, edited by B. Hewlett and M. Lamb. pp. 147-171. Somerset, NJ: Transaction Publishers.

Tullberg, Birgitta S., and Virpi Lummaa (2002). Induced abortion ratio in modern Sweden falls with age, but rises again before menopause. *Evolution and Human Behavior* 22(1): 1-10.

Tymicki, Krzysztof (2004). Kin influence on female reproductive behavior: The evidence from reconstitution of the Bejsce parish registers, 18th to 20th centuries, Poland. *American Journal of Human Biology* 16(5): 508-522.

U.S. Department of the Interior (1972). *Proposed Obed National Wild and Scenic River, Tennessee* . Draft Environmental Statement. Copy of report housed at the Fish and Wildlife Service office in Cookeville, Tennessee.

Ugan, A., and J. Bright (2002). Measuring foraging efficiency with archaeological faunas: Exploring the relationship between abundance indices and foraging returns. *Journal of Archaeological Science* 28: 1309-1321.

Valeggia, Claudia, and Peter T. Ellison (2003). Energetics, fecundity, and human life history. In *The Biodemography of Human Reproduction and Fertility*, J. Rodgers and H. P. Kohler, eds., pp. 87-103. Boston: Kluwer Academic Publishers.

Voland, Eckart and Athanasios Chasiotis and Wulf Schiefenhovel (2005). Introduction: Grandmotherhood: A short overview of three fields of research on the evolutionary significance of postgenerative female life. In *Grandmotherhood: The Evolutionary Significance of the Second Half of the Female Life*. Edited by E. Voland, C. Athanasios, and W. Schiefenhovel. pp 1-20. Piscataway: Rutgers University Press.

Voland, Eckart, and Jan Beise (2002). Opposite effects of maternal and paternal grandmothers on infant survival in historical Krummhörn. *Behavioral Ecology and Sociobiology* 52: 435-443.

Voland, Eckart, and Jan Beise (2005). "The husband's mother is the devil in house" - Data on the impact of the mother-in-law on stillbirth mortality in historical Krummhörn (c18-c19 Germany) and some thoughts on the evolution of postgenerative female life. In Voland, Eckart; Chasiotis, Athanasios & Schiefenhövel, Wulf (eds.): *Grandmotherhood: The Evolutionary Significance of the Second Half of Female Life*. pp. 239-255 New Brunswick & London (Rutgers University Press).

Walker, G., Parisher, E., Smith, P., Whitlock, D., Kramer, D., Matthes, U., and Morefield, L. (n.d.). *Characterization of plant community structure and abiotic conditions on climbed and unclimbed cliff faces in the Obed River Gorge* . Copy of report housed at the Fish and Wildlife Service office in Cookeville, Tennessee.

Waynforth, David (2002). Evolutionary theory and reproductive responses to father absence: Implications of kin selection and the reproductive returns to mating and parenting effort. In *Handbook of Father Involvement*. Eds. C. Tamis-LeMonda and N. Cabrera. Lawrence Erlbaum Associates, New Jersey.

Williams, C.E., Bivens, R.D., and Carter, B.D. (2003, May). *A crayfish survey on Clear Creek, Morgan County, Tennessee* . Tennessee Wildlife Resources Agency.

Wilson, David Sloan (2005). Testing major evolutionary hypotheses about religion with a random sample. *Human Nature* 16(4): 382-409.

Winterhalder, B (2002). Intra-group resource transfers: Comparative evidence, models, and implications for human evolution. In *Meat Eating and Human Evolution*, C.B. Stanford and H.T. Bunn, eds., pp. 279-301. Oxford: Oxford University Press.

Winterhalder, B (2002). The behavioural ecology of hunter-gatherers. In *Hunter-Gatherers: An Interdisciplinary Perspective*. Catherine Panter-Brick, Robert H.

Layton and Peter Rowley-Conwy, eds., pp. 12-38. Cambridge: Cambridge University Press.

Wolf, Arthur P (2003). Maternal sentiments: How strong are they? *Current Anthropology* 44(S5): S31-S50.

Wolfe, W.J. (2007, April). Geomorphology and vegetation of alluvial bars along the Obed Wild and Scenic River . *Proceedings of the Seventeenth Tennessee Water Resources Symposium*, Montgomery Bell State Park, Burns, Tennessee.

Wood, Brian (2006). Prestige or provisioning? A test of foraging goals among the Hadza. *Current Anthropology* 47(2):383-387.

Worthman CM, and Brown RA (2005). A biocultural life history approach to the developmental psychobiology of male aggression. In DM Stoff, EJ Susman (eds) *Developmental Psychobiology of Aggression*. pp. 187-222. Cambridge: Cambridge University Press.

Worthman, Carol M., and Jennifer Kuzara (2005). Life history and the early origins of health differentials. *American Journal of Human Biology* 17(1): 95-112.

Yanca, Catherine, and Bobbi S. Low (2004). Female allies and female power: A cross-cultural analysis. *Evolution and Human Behavior* 25(1): 9-23.

Yang, Zhengwei and Jeffrey C. Schank (2006). Women do not synchronize their menstrual cycles. *Human Nature* 17(4): 434-447.

Index